BARRON'S

Regents Exams and Answers

Chemistry—Physical

Setting Revised Edition

ALBERT S. TARENDASH
Assistant Principal—Supervision (Retired), Department of Chemistry and Physics Stuyvesant High School, New York, New York

KATE GEFELL
Chemistry Teacher, New York State Master Teacher—Chemistry, Southern Tier Region
Ithaca High School, Ithaca, New York

Published by Kaplan North America, LLC d/b/a Barron's Educational Series
1515 West Cypress Creek Road
Fort Lauderdale, Florida 33309
www.barronseduc.com

ISBN: 978-1-5062-6468-4

10 9 8 7 6 5 4

Kaplan North America, LLC d/b/a Barron's Educational Series print books are available at special quantity discounts to use for sales promotions, employee premiums, or educational purposes. For more information or to purchase books, please call the Simon & Schuster special sales department at 866-506-1949.

Contents

Regents Examinations, Answers, and
Self-Analysis Charts   83

Preface

A HELPFUL WORD TO THE STUDENT

As you are aware, the purpose of this book is to help you review for the New York State Regents examination in Chemistry—Physical Setting. You can also use it effectively to prepare for classroom, midterm, and final examinations. The book contains a number of special sections and other features designed to aid you in achieving good grades on your examinations. Included are the following:

- *How to Use This Book.* Your journey begins here. This section explains how to use the *entire* book as an effective test-preparation tool. It provides a method of identifying those areas you have mastered and those that will require additional work.
- *Test-Taking Techniques.* In this section, you will learn *how* to take the Regents examination effectively. Included are a list of materials you will need to bring to the examination and directions on how to fill out your answer sheet, how to read the questions, and how to deal with difficult items. In addition, it provides some suggestions on how to approach the extended-response questions found on Part C of the examination.
- *What to Expect on the Chemistry Examination.* This section explains the format, content, and grading of the examination.
- *New York State Physical Setting/Chemistry Core: Topic Outline and Question Index.* The topic outline provides a detailed description of the Regents Physical Setting/Chemistry Core, and the question index keys the Regents examination questions that appear in this book to the topic outline. The index will help you find questions on similar topics and will provide information about the areas that are stressed most heavily on the exam.
- *Glossary of Important Terms.* The glossary will help you understand the technical terms that may appear on the examination.
- *Reference Tables for Chemistry.* A significant part of the Regents examination requires you to be able to find and use the information contained in these reference tables. This section describes each table and the types of information that may be obtained from it.

- *Recent Regents Examinations.* These questions will provide the bulk of your study and review. When you have answered *all* of the questions that appear in this book, you will be able to approach the Regents examination with confidence!
- *Explanation of Answers and Self-Analysis Chart for Each Examination.* The detailed answer explanations will help you to *understand* why certain choices are correct, and others are incorrect. The self-analysis chart will pinpoint your strengths and weaknesses on each practice examination you take.

This book was written to provide you with the basic tools you need to perform well on the Regents Chemistry—Physical Setting examination. If you follow its advice and prepare diligently, you will achieve your goal.

Best wishes for success!

Albert S. Tarendash

How to Use This Book

1. Read the section entitled *Test-Taking Techniques* to learn how to prepare properly for an examination and how to take it with maximum efficiency.
2. Read the section entitled *What to Expect on the Chemistry Examination* to familiarize yourself with the structure and contents of this examination.
3. Read the section entitled *Reference Tables for Chemistry* to familiarize yourself with the contents and use of these tables.
   Note: On occasion, the New York State Education Department may change the content or format of the Regents examination and/or the reference tables. Your classroom teacher is your best source of information about such changes.
4. Read the section entitled *Using the Equations to Solve Chemistry Problems*.
5. Refer to the *Glossary of Important Terms* to learn the meanings of words and terms you do not understand.
6. Take the first Regents examination in this book, answering all of the questions.
7. Check your answers, and then complete the self-analysis chart at the end of the examination to pinpoint your strengths and weaknesses.
8. Read the detailed explanation of *all of the questions,* paying closest attention to the questions you answered incorrectly. Occasionally, the *wrong choices* are explained, and these explanations may help you understand why you chose an incorrect answer.
9. When you have determined your areas of weakness, refer to the *New York State Physical Setting/Chemistry Core: Topic Outline and Question Index* to locate similar questions on other recent examinations. (You can also use this outline and index to determine which areas have been stressed in recent years.)
10. Repeat steps 6–9 for the other examinations, *with the exception of the most recent test.*

11. When you have completed your studying, but no more than 1 or 2 days before the actual examination, take the most recent examination in this book *under strict examination conditions.*

After you have checked your answers to this last examination, you will have a rough idea of how you will perform on the Regents examination you will take.

Test-Taking Techniques

HELPFUL HINTS

The following pages contain 8 tips to help you achieve a good grade on the Chemistry Regents examination.

TIP 1
Be confident and prepared.

SUGGESTIONS

- Review previous tests.
- Use a clock or watch and take the most recent exam at home under examination conditions (i.e., don't have the radio or television on).
- Get a review book. (One useful book is Barron's *Let's Review Regents: Chemistry—Physical Setting*.)
- Talk over the answers to questions on these tests with someone else, such as another student in your class or someone at home.
- Finish all your homework assignments.
- Look over classroom exams that your teacher gave during the term.
- Take class notes carefully.
- Practice good study habits.
- Know that there are answers for every question.
- Be aware that the people who made up the Regents examination want you to succeed.
- Remember that thousands of students over the last few years have taken and passed a Chemistry Regents. You can pass too!
- Complete your study and review at least one day before the examination. Last-minute cramming does not help and may hurt your performance.

- On the night prior to the exam day, lay out all the things you will need.
- Go to bed early; eat wisely.
- Bring the required materials to the examination. This generally means a pen, two sharpened pencils, and a good quality eraser. If your school does not supply a calculator, be certain to bring one to the examination. Some schools require a signed Regents admission card for identification. Good advice: Assume your school will *not* supply you with any materials!
- Once you are in the exam room, arrange things, get comfortable, be relaxed, attend to personal needs (the bathroom).
- Keep your eyes on your own paper; do not let them wander over to anyone else's paper.
- Be polite in making any reasonable demands of the exam-room proctor, such as changing your seat or having window shades raised or lowered.

TIP 2

Read test instructions carefully.

SUGGESTIONS

- Be familiar with the format of the examination.
- Know how the test will be graded.
- If your school supplies an electronic scoring sheet, be certain you are familiar with the additional directions for recording and changing answers.
- If you decide to change an answer, be certain that you erase your original response completely.
- Any stray marks on your answer sheet should be erased completely.
- Be familiar with the directions for Parts B and C questions. Answer each question completely. Explanations must be written as *whole sentences* and substitutions into equations *must* include units. Be certain that your answers are clearly labeled and well organized. Place a box around numerical answers. Be neat!
- Ask for assistance from the exam-room proctor if you do not understand the directions.

TIP 3

Read each question carefully and read each choice before you record your answer.

SUGGESTIONS

- Be sure you understand *what* the question is asking.
- Try to recognize information that is *given* in the question.
- Will a chemistry reference table help you find the answer to the question?
- Some choices may look appealing yet will be incorrect.
- Try to eliminate those choices that are *obviously* incorrect.

TIP 4

Budget your test time (3 hours).

SUGGESTIONS

- Bring a watch or clock to the test.
- The Regents examination is designed to be completed in $1\frac{1}{2}$ to 2 hours.
- If you are absolutely uncertain of the answer to a question, mark your question booklet and move on to the next question.
- If you persist in trying to answer every difficult question *immediately*, you may find yourself rushing or unable to finish the remainder of the examination.
- When you have finished the examination, return to those unanswered questions.
- Good advice: If at all possible, reread the *entire* examination—and your responses—at least one more time. (This will help you eliminate those errors that result from misreading questions.)

TIP 5

Use your reasoning skills.

SUGGESTIONS

- Answer *all* questions.
- Relate (connect) the question to anything that you studied, wrote in your notebook, or heard your teacher say in class.
- Relate (connect) the question to any film, demonstration, or experiment you saw in class, any project you did, or to anything you may have learned from newspapers, magazines, or television.
- Look over the entire test to see whether one part of it can help you answer another part.

TIP 6

Use your reference tables.

SUGGESTIONS

- You should be familiar with the *content* of each table.
- Frequently, the answers to questions can be found from information contained within these tables.
- Some questions refer to specific tables.
- Other questions do not refer to a table, but can be answered by choosing and using the correct table.
- Be especially familiar with the Periodic Table of the Elements, as well as Reference Table *T*, Important Formulas and Equations.

TIP 7

Don't be afraid to guess.

SUGGESTIONS

- Eliminate obvious incorrect choices.
- If still unsure of an answer, make an educated guess.
- There is no penalty for guessing; therefore, answer *all* questions. An omitted answer gets no credit.

TIP 8
Sign the declaration.

SUGGESTIONS

- Be certain that you sign the declaration found on your answer sheet.
- Unless this declaration is signed, your paper cannot be scored.

SUMMARY OF TIPS

1. Be confident and prepared.
2. Read test instructions carefully.
3. Read each question carefully and read each choice before you record your answer.
4. Budget your test time (3 hours).
5. Use your reasoning skills.
6. Use your reference tables.
7. Don't be afraid to guess.
8. Sign the declaration.

HOW TO ANSWER PART C (EXTENDED-RESPONSE) QUESTIONS

An *extended-response question* is an examination question that requires the test taker to do more than choose among several responses or fill in a blank. You may need to perform numerical calculations, draw and interpret graphs, and provide extended written responses to a question or a problem.

Part C of the New York State Regents Examination in Chemistry—Physical Setting contains extended-response questions. This section is designed to provide you with a number of general guidelines for answering them.

SOLVING PROBLEMS INVOLVING NUMERICAL CALCULATIONS

To receive full credit you must:

- Provide the appropriate equation(s).
- Substitute values and units into the equation(s).
- Display the answer, with appropriate units and to the correct number of significant figures.
- If the answer is a vector quantity, include its direction.

You should write as legibly as possible. Teachers are human, and nothing irks them more than trying to decipher a careless, messy scrawl. It is also a good idea to identify your answer clearly, either by placing it in a box or by writing the word "answer" next to it.

A final word: If you provide the correct answer but do not show any work, you will not receive any credit for the problem!

The following is a sample problem and its model solution.

Problem

A 5.00-gram object has a density of 4.00 grams per cubic centimeter. Calculate the volume of this object.

Solution

$$d = \frac{m}{V}$$

Rearranging the equation gives

$$V = \frac{m}{d} = \frac{5.00 \text{ g}}{4.00 \text{ g/cm}^3}$$

$$\boxed{V = 1.25 \text{ cm}^3}$$

GRAPHING EXPERIMENTAL DATA

To receive full credit you must:

- Label both axes with the appropriate variables and units.
- Divide the axes so that the data ranges fill the graph as nearly as possible.
- Plot all data points accurately.
- Draw a best-fit line carefully with a straightedge. The line should pass through the origin *only if the data warrant it.*
- If a part of the question requires that the slope be calculated, calculate the slope *from the line,* not from individual data points.

A graph should have a title, and the *independent variable* is usually drawn along the x-axis.

The following is a sample problem and its model solution.

Problem

A student attempts to estimate absolute zero in the following way: He subjects a sample of gas (at constant pressure and mass) to varying temperatures and measures the gas volume at each of the temperatures. The accompanying table contains his experimental data.

Temperature/°C	Volume/mL
−100	128
−60	148
−40	156
0	204
40	222
80	272
160	310

(a) Using axes that are appropriately labeled and scaled, draw a graph that accurately displays the student's data.
(b) Estimate the student's value for absolute zero by extending the graph to the Celsius temperature at which the volume of the gas is 0 milliliter.

Solution

(a) The first step is to construct an appropriate set of axes if one is not provided. We will assume that you must start from scratch. Since temperature is the independent variable, you need to place it along the x-axis. Also, since absolute zero is $-273.15°C$, you must scale the axes properly. Here is one appropriate set of axes:

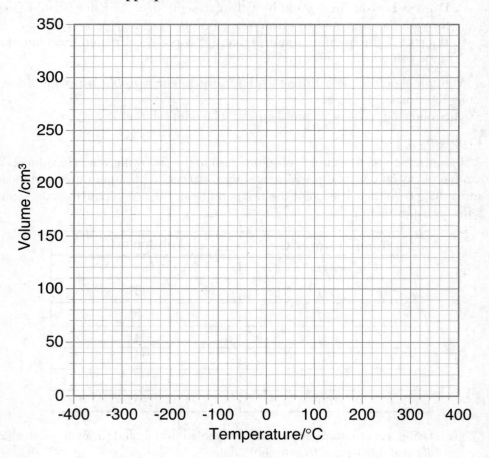

Second, you must plot the data points carefully on the axes as shown below:

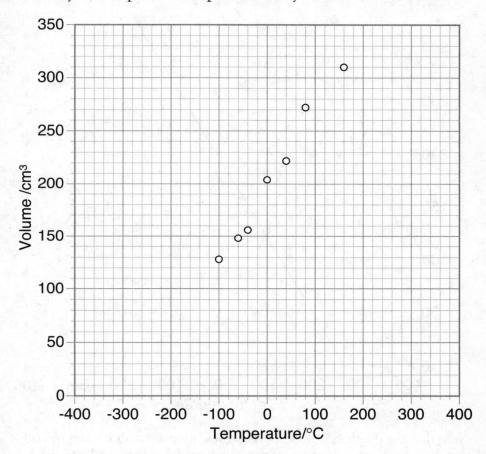

Your third task is to draw the graph. You might be tempted to "connect the dots," but then you would miss the significant relationship between volume and temperature. If you examine *all* of the plotted points, you will note that they fall approximately on a *straight line*.

Therefore, the next step is to draw a *best-fit* straight-line graph. This is a graph in which the data points are most closely distributed on both sides of the line. The accompanying graph shows the best-fit straight line.

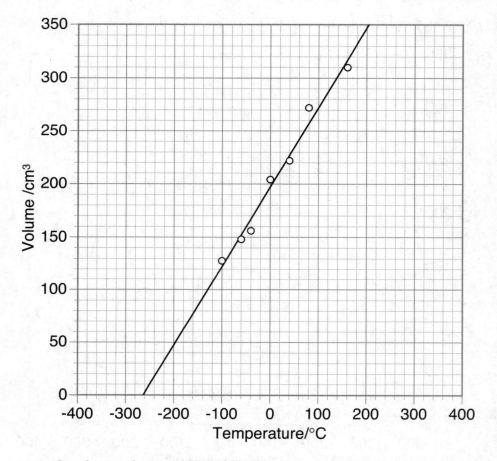

Note that the graph extending beyond the data points (above and below) is a *dashed* line. Such extensions are known as *extrapolated data*, and they are based on the assumption that the gas will continue to behave as it did within the experimental range for which the student collected data.

(You may ask: Why *this* particular line? It seems as though many lines could have been drawn using the plotted data. Actually, there is only *one* best-fit straight line, and it is calculated by using an advanced statistical technique known as *linear regression*. This technique was used to draw the line shown above. At this point, it is sufficient for you to provide an "eyeball" estimate of the best-fit straight line.)

   (b) Your final task is to inspect the graph closely and to estimate absolute zero. The calculated value is −263°C. (This corresponds to an experimental error of 3.7%.)

DRAWING DIAGRAMS

To receive full credit you must:

- Draw your diagrams neatly and label them clearly.
- Bring a straightedge and a protractor with you so that you can draw neat, accurate diagrams.

The accompanying diagram represents a zinc–copper electrochemical cell containing an agar–KCl salt bridge. This is the type of diagram you may be asked to draw as part of an examination question.

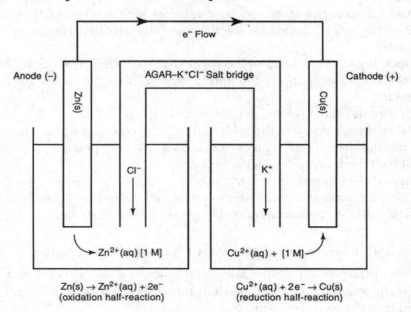

$Zn(s) \rightarrow Zn^{2+}(aq) + 2e^-$
(oxidation half-reaction)

$Cu^{2+}(aq) + 2e^- \rightarrow Cu(s)$
(reduction half-reaction)

WRITING A FREE-RESPONSE ANSWER

To receive full credit you must:

- Use complete, clear sentences that make sense to the reader.
- Use correct chemistry in your explanations.

A sample question and a model answer follow.

Question

Describe in detail the technique used to determine the concentration of a dilute hydrochloric acid solution using a dilute sodium hydroxide solution of known concentration.

Name any equipment or other chemicals that you would use. *You need not describe any mathematical calculations.*

Model Answer

Since this question requires an extended answer containing a series of steps, we decide to use an outline form.

The technique is called *titration*, and it is described in the steps given below.

- Place a known volume of the hydrochloric acid solution in a beaker of suitable size.
- Add a drop or two of an acid–base indicator such as phenolphthalein.
- Pour the sodium hydroxide solution carefully into a burette, using a small funnel.
- Open the stopcock to allow the trapped air to escape. Then close the stopcock and wipe the tip of the burette with a tissue to remove any clinging liquid.
- Record the initial volume in the burette.
- Add the base slowly to the acid solution, with continuous stirring, until the phenolphthalein just changes from colorless to faint pink.
- Record the final volume in the burette.
- Repeat the experiment at least once.
- Rinse the apparatus with water to remove all traces of acid and base.
- Calculate the concentration of the hydrochloric acid solution.

Writing and Balancing Equations in a Chemistry Examination

For all of the suggestions that follow, the reaction that occurs between aqueous solutions of sodium sulfate and barium nitrate is used as an example and is referred to as the given reaction.

- If you are asked to write a *word equation*, be certain to include the correct names of the reactants and products and their phases in the reaction.

For the given reaction, the word equation is

$$\text{barium nitrate(aq)} + \text{sodium sulfate(aq)} \rightarrow$$
$$\text{barium sulfate(s)} + \text{sodium nitrate(aq)}$$

- If you are asked to write a *balanced* equation, you are usually expected to balance using *smallest whole-number coefficients*.

  For the given reaction, the balanced equation is

  $$Ba(NO_3)_2\,(aq) + Na_2SO_4\,(aq) \rightarrow BaSO_4\,(s) + 2NaNO_3\,(aq)$$

- If you are asked to write an *ionic* equation occurring in aqueous solution, you must reduce everything to its component ions except *insoluble* compounds, such as $BaSO_4(s)$, and (of course!) *covalently* bonded substances, such as $H_2O(\ell)$.

  For the given reaction, the ionic equation is

  $$Ba^{2+}(aq) + 2NO_3^-(aq) + 2Na^+(aq) + SO_4^{2-} \rightarrow$$
  $$BaSO_4(s) + 2Na^+(aq) + 2NO_3^-(aq)$$

- If you are asked to write a *net* ionic equation, you should omit all *spectator ions*, that is, all ions appearing *unchanged on both sides of the equation*.

  For the given reaction, the net ionic equation is

  $$Ba^{2+}(aq) + SO_4^{2-}(aq) \rightarrow BaSO_4(s)$$

What to Expect on the Chemistry Examination

FORMAT OF THE CHEMISTRY EXAMINATION

The chemistry examination will be 3 hours long and will include three parts: A, B, and C. You should be prepared to answer questions in multiple-choice format, as well as answer questions that require a more extended response.

Questions will be content- and skills-based, and you may be required to graph data, complete a data table, label or draw diagrams, design experiments, make calculations, or write short or extended responses.

In addition, you may be required to hypothesize, interpret, analyze, evaluate data, or apply your scientific knowledge and skills to real-world situations.

[In the future, a Part D will be added, which will focus on assessment of laboratory skills. As more information becomes available, the New York State Education Department will inform schools of the development status of the performance test.]

You will be required to answer ALL of the questions on the Physical Setting/Chemistry Regents examination.

Physical Setting/Chemistry Regents Examination Format

PART	ITEM TYPE(S)	DESCRIPTION OF THE ITEMS	APPROXIMATE PERCENT OF TOTAL TEST RAW SCORE
A	Multiple-choice questions	Content-based questions assessing your knowledge and understanding of core material	35
B	Multiple-choice and constructed-response questions	Content- and skills-based questions assessing your ability to apply, analyze, and evaluate material	30
C	Constructed-response and/or extended constructed-response questions	Content-based and application questions assessing your ability to apply knowledge of science concepts and skills to address real-world situations	20

The maximum *raw* score on the examination is 85 points. A teacher's chart will be provided for converting your *raw* score to a *scaled* score that has a maximum of 100 points. A sample conversion table taken from the June 2004 Regents Chemistry—Physical Setting examination is shown below:

Sample Conversion Table

Raw Score	Scaled Score	Raw Score	Scaled Score	Raw Score	Scaled Score	Raw Score	Scaled Score
85	100	63	74	41	58	19	35
84	98	62	73	40	57	18	34
83	97	61	73	39	56	17	33
82	95	60	72	38	55	16	31
81	94	59	71	37	55	15	30
80	93	58	70	36	54	14	28
79	91	57	69	35	53	13	26
78	90	56	69	34	52	12	25
77	89	55	68	33	51	11	23
76	87	54	67	32	50	10	21
75	86	53	66	31	49	9	19
74	85	52	66	30	48	8	17
73	84	51	65	29	47	7	15
72	83	50	64	28	46	6	13
71	82	49	64	27	45	5	11
70	81	48	63	26	44	4	9
69	80	47	62	25	43	3	7
68	79	46	61	24	42	2	5
67	78	45	61	23	41	1	2
66	77	44	60	22	39	0	0
65	76	43	59	21	38		
64	75	42	58	20	37		

The table is used to convert the number of points you actually received on the examination (your "raw" score) to your final score on the examination (your "scaled" score). **Note that this table will change from one examination to another.**

TOPICS COVERED ON THE CHEMISTRY EXAMINATION

All of the questions on the Chemistry examination will test major understandings, skills, and real-world applications drawn from the following 12 subject areas:

M.	Math Skills
R.	Reading Skills
I.	Atomic Concepts
II.	Periodic Table
III.	Moles/Stoichiometry
IV.	Chemical Bonding
V.	Physical Behavior of Matter
VI.	Kinetics/Equilibrium
VII.	Organic Chemistry
VIII.	Oxidation–Reduction
IX.	Acids, Bases, and Salts
X.	Nuclear Chemistry

It is suggested that you read the Topic Outline found on pages 20–38 in order to learn the exact nature of the material that is subject to testing.

New York State Physical Setting/ Chemistry Core

TOPIC OUTLINE

The Topic Outline on pages 20–38 is adapted from Appendix B of the New York State Physical Setting/Chemistry Core. All Regents Chemistry—Physical Setting Examinations are based on this core. The topic outline is divided into 12 sections:

M.	Math Skills
R.	Reading Skills
I.	Atomic Concepts
II.	Periodic Table
III.	Moles/Stoichiometry
IV.	Chemical Bonding
V.	Physical Behavior of Matter
VI.	Kinetics/Equilibrium
VII.	Organic Chemistry
VIII.	Oxidation–Reduction
IX.	Acids, Bases, and Salts
X.	Nuclear Chemistry

Each section contains one or more of the following items:

- The *Major Understandings* that you must have mastered for the examination
- The *Skills* that you need to be able to demonstrate during the examination
- The *Key Points to Remember* that remind you of key chemistry concepts that you should be familiar with to succeed on this examination.

M. MATHEMATICS SKILLS NEEDED FOR CHEMISTRY

M.1 Organize, graph, and analyze data gathered from laboratory activities or other sources.

- Identify independent and dependent variables.
- Create appropriate axes with label and scale.
- Identify graph points clearly.

M.2 Interpret a graph constructed from experimentally determined data.

- Identify direct and inverse relationships.
- Apply data showing trends to predict information.

M.3 Measure and record experimental data and use the data in calculations.

- Choose appropriate measurement scales and use units in recording.
- Show mathematical work stating formula and steps for solution.
- Estimate answers.
- Use appropriate equations and significant digits.
- Identify relationships within variables from data tables.
- Calculate percent error.

M.4 Recognize and convert various scales of measurement.

- Convert between Celsius (°C) and Kelvin (K).
- Convert among kilometers (km), meters (m), centimeters (cm), and millimeters (mm).
- Convert between grams (g) and kilograms (kg).
- Convert between kilopascals (kPa) and atmospheres (atm).

M.5 Employ critical thinking skills in solving problems.

- Apply algebraic or geometric concepts in the solution of mathematical problems.
- Use knowledge of geometric arrangements to predict particle properties or behavior.
- State the assumptions on which a particular mathematical equation is based.
- Evaluate the appropriateness of an answer to a solved problem.

R. READING SKILLS NEEDED FOR CHEMISTRY

R.1 Reading Comprehension: Literal

- Extract an answer from the text when the answer is "right there" in the information provided.
- Read and interpret information provided in the Reference Tables for Chemistry.

R.2 Reading Comprehension: Inferential/Interpretive

- Understand and make connections between prior content knowledge and information from the text when these connections are not explicitly stated.
- Understand and apply new information provided in the text and solve problems using this information.

R.3 Reading Comprehension: Lexical

- Understand and apply key vocabulary in a reading passage.

I. Atomic Concepts

	MAJOR UNDERSTANDINGS	SKILLS The student should be able to:	KEY POINTS TO REMEMBER
I.1	The modern model of the atom has evolved over a long period of time through the work of many scientists.	relate experimental evidence to models of the atom	Examples: 1. J.J. Thomson and cathode ray experiment 2. Ernest Rutherford and gold foil experiment
I.2	Each atom has a nucleus, with an overall positive charge, surrounded by negatively charged electrons.	use models to describe the structure of an atom	
I.3	Subatomic particles contained in the nucleus include protons and neutrons.		
I.4	The proton is positively charged, and the neutron has no charge. The electron is negatively charged.		
I.5	Protons and electrons have equal but opposite charges. The number of protons is equal to the number of electrons in an atom.	determine the number of protons or electrons in an atom or ion when given one of these values	Positive ions form when atoms lose electrons. Negative ions form when atoms gain electrons.
I.6	The mass of each proton and each neutron is approximately equal to one atomic mass unit. An electron is much less massive than a proton or neutron.	calculate the mass of an atom, the number of neutrons, or the number of protons, given the other two values	
I.7	In the wave-mechanical model (electron cloud), the electrons are in orbitals, which are defined as regions of most probable electron location (ground state).	interpret electron configurations listed on the Periodic Table	
I.8	Each electron in an atom has its own distinct amount of energy.		Electron configurations begin with the lowest energy shell near the nucleus and continue in order of increasing energy and distance from the nucleus.
I.9	When an electron in an atom gains a specific amount of energy, the electron is at a higher energy state (excited state).	distinguish between ground state and excited state electron configurations, e.g., 2−8−2 vs. 2−7−3	Excited state configurations contain the same number of electrons as ground state configurations.

I. Atomic Concepts (*Continued*)

	MAJOR UNDERSTANDINGS	SKILLS The student should be able to:	KEY POINTS TO REMEMBER
I.10	When an electron returns from a higher energy state to a lower energy state, a specific amount of energy is emitted. This emitted energy can be used to identify an element.	identify an element by comparing its bright-line spectrum to given spectra	
I.11	The outermost electrons in an atom are called the valence electrons. In general, the number of valence electrons affects the chemical properties of an element.	draw a Lewis electron-dot structure of an atom distinguish between valence and non-valence electrons, given an electron configuration, e.g., 2–8–2	
I.12	Atoms of an element that contain the same number of protons but a different number of neutrons are called isotopes of that element.		
I.13	The average atomic mass of an element is the weighted average of the masses of its naturally occurring isotopes.	given an atomic mass, determine the most abundant isotope calculate the atomic mass of an element, given the masses and ratios of naturally occurring isotopes	

II. Periodic Table

	MAJOR UNDERSTANDINGS	SKILLS The student should be able to:	KEY POINTS TO REMEMBER
II.1	The placement or location of an element on the Periodic Table gives an indication of physical and chemical properties of that element. The elements on the Periodic Table are arranged in order of increasing atomic number.	explain the placement of an unknown element in the Periodic Table based on its properties	
II.2	The number of protons in an atom (atomic number) identifies the element. The sum of the protons and neutrons in an atom (mass number) identifies an isotope. Common notations that represent isotopes include: ^{14}C, $^{14}_{6}C$, carbon-14, C-14.	interpret and write isotopic notation	Subtract the atomic number from the mass number to find the number of neutrons in an isotope! The term "nuclear charge" refers to the number of protons in an atom's nucleus. For example, a carbon atom with 6 protons would have a nuclear charge of 6+.

II. Periodic Table (*Continued*)

	MAJOR UNDERSTANDINGS	**SKILLS** The student should be able to:	**KEY POINTS TO REMEMBER**
II.3	Elements can be classified by their properties and located on the Periodic Table as metals, nonmetals, metalloids (B, Si, Ge, As, Sb, Te), and noble gases.	classify elements as metals, nonmetals, metalloids, or noble gases by their properties	
II.4	Elements can be differentiated by their physical properties. Physical properties of substances, such as density, conductivity, malleability, solubility, and hardness, differ among elements.	describe the states of the elements at STP suggest a simple separation strategy to separate two or more elements based on physical properties	
II.5	Elements can be differentiated by chemical properties. Chemical properties describe how an element behaves during a chemical reaction.		Metal atoms lose electrons when they bond.
II.6	Some elements exist as two or more forms in the same phase. These forms differ in their molecular or crystal structure and hence in their properties.		Allotropes of carbon include graphite, diamond, and fullerene.
II.7	For Groups 1, 2, and 13–18 on the Periodic Table, elements within the same group have the same number of valence electrons (helium is an exception) and, therefore, similar chemical properties.	determine the group of an element, given the chemical formula of a compound, e.g., XCl or XCl_2	
II.8	The succession of elements within the same group demonstrates characteristic trends: differences in atomic radius, ionic radius, electronegativity, first ionization energy, and metallic/nonmetallic properties.	compare and contrast properties of elements within a group or a period for Groups 1, 2, 13–18 on the Periodic Table	Increasing the distance between the nucleus and an atom's valence electrons causes the ionization energy and the electronegativity to decrease down each group of the Periodic Table.
II.9	The succession of elements across the same period demonstrates characteristic trends: differences in atomic radius, ionic radius, electronegativity, first ionization energy, and metallic/nonmetallic properties.		Increasing nuclear charge causes electronegativity and ionization energy to increase across each period of the Periodic Table.

III. Moles/Stoichiometry

	MAJOR UNDERSTANDINGS	SKILLS The student should be able to:	KEY POINTS TO REMEMBER
III.1	A compound is a substance composed of two or more different elements that are chemically combined in a fixed proportion. A chemical compound can be broken down by chemical means. A chemical compound can be represented by a specific chemical formula and assigned a name based on the IUPAC system.	determine how many total atoms are in a compound determine how many atoms of each type are in a compound use IUPAC nomenclature rules to name binary ionic, binary molecular, and polyatomic ionic compounds	Compounds, along with elements, are pure substances.
III.2	Types of chemical formulas include empirical, molecular, and structural.	identify examples of empirical, structural, or molecular formulas convert a structural formula into a molecular formula or an empirical formula	
III.3	The empirical formula of a compound is the simplest whole-number ratio of atoms of the elements in a compound. It may be different from the molecular formula, which is the actual ratio of atoms in a molecule of that compound.	determine the molecular formula, given the empirical formula and molecular mass determine the empirical formula from a molecular formula	
III.4	In all chemical reactions there is a conservation of mass, energy, and charge.	interpret balanced chemical equations in terms of conservation of matter and energy	Atoms In = Atoms Out!
III.5	A balanced chemical equation represents conservation of atoms. The coefficients in a balanced chemical equation can be used to determine mole ratios in the reaction.	balance equations, given the formulas for reactants and products interpret balanced chemical equations in terms of conservation of matter and energy create and use models of particles to demonstrate balanced equations calculate simple mole—mole stoichiometry problems, given a balanced equation	Coefficients in a balanced chemical equation represent moles or particles, *not* grams!
III.6	The formula mass of a substance is the sum of the atomic masses of its atoms. The molar mass (gram-formula mass) of a substance equals one mole of that substance.	calculate the formula mass and the gram-formula mass use the molar mass to convert from grams to moles or from moles to grams	Formula masses are reported in atomic mass units (amu). Molar masses are reported in grams/mol.

III. Moles/Stoichiometry (*Continued*)

		MAJOR UNDERSTANDINGS	SKILLS The student should be able to:	KEY POINTS TO REMEMBER
III.7		The percent composition by mass of each element in a compound can be calculated mathematically.	determine the number of moles of a substance, given its mass determine the mass of a given number of moles of a substance use lab data to determine the percent of water in a hydrate	
III.8		Types of chemical reactions include synthesis, decomposition, single replacement, and double replacement.	identify types of chemical reactions	Synthesis reactions create just one product: $$A + B \rightarrow AB$$ Decomposition reactions have just a single reactant: $$AB \rightarrow A + B$$ Single replacement reactions have uncombined elements on *both* sides of the reaction: $$A + BC \rightarrow AC + B$$ Double replacement reactions show "partners switching places": $$AB + CD \rightarrow AD + CB$$

IV. Chemical Bonding

		MAJOR UNDERSTANDINGS	SKILLS The student should be able to:	KEY POINTS TO REMEMBER
IV.1		Compounds can be differentiated by their chemical and physical properties.	distinguish among ionic, molecular, and metallic substances, given their properties	
IV.2		Two major categories of compounds are ionic and molecular (covalent) compounds.		
IV.3		Chemical bonds are formed when valence electrons are: transferred from one atom to another (ionic); shared between atoms (covalent); mobile within a metal (metallic).	demonstrate bonding concepts using Lewis dot structures representing valence electrons: transferred (ionic bonding); shared (covalent bonding); in a stable octet	Metallic bonds hold samples of pure metals together. Ionic bonds form between metal ions and nonmetal ions. Covalent bonds form between nonmetal or metalloid atoms.

IV. Chemical Bonding (*Continued*)

	MAJOR UNDERSTANDINGS	SKILLS The student should be able to:	KEY POINTS TO REMEMBER
IV.4	In a multiple covalent bond, more than one pair of electrons are shared between two atoms. Unsaturated organic compounds contain at least one double or triple bond.	evaluate the number of electrons shared, or the number of electron pairs shared, in single, double, or triple covalent bonds	
IV.5	Molecular polarity can be determined by the shape and distribution of the charge. Symmetrical (nonpolar) molecules include CO_2, CH_4, and diatomic elements. Asymmetrical (polar) molecules include HCl, NH_3, H_2O.		Molecular symmetry cancels polarity! Draw Lewis electron-dot diagrams to determine molecular shape.
IV.6	When an atom gains one or more electrons, it becomes a negative ion and its radius increases. When an atom loses one or more electrons, it becomes a positive ion and its radius decreases.		
IV.7	When a bond is broken, energy is absorbed. When a bond is formed, energy is released.		
IV.8	Atoms attain a stable valence electron configuration by bonding with other atoms. Noble gases have stable valence electron configurations and tend not to bond.	determine the noble gas configuration an atom will achieve when bonding	Remember the octet rule! Atoms (except hydrogen) tend to bond so as to achieve the stable octet of eight valence electrons.
IV.9	Physical properties of substances can be explained in terms of chemical bonds and intermolecular forces. These properties include conductivity, malleability, solubility, hardness, melting point, and boiling point.	compare and contrast the physical properties of different compounds based on the types of chemical bonds and/or intermolecular forces they have	
IV.10	Electron-dot diagrams (Lewis structures) can represent the valence electron arrangement in elements, compounds, and ions.	demonstrate bonding concepts, using Lewis dot structures representing valence electrons: transferred (ionic bonding); shared (covalent bonding); in a stable octet	Check! Does your drawing show all of the valence electrons? Does each atom (except H) have 8 valence electrons?

IV. Chemical Bonding (*Continued*)

	MAJOR UNDERSTANDINGS	**SKILLS** The student should be able to:	**KEY POINTS TO REMEMBER**
IV.11	Electronegativity indicates how strongly an atom of an element attracts electrons in a chemical bond. Electronegativity values are assigned according to arbitrary scales.		
IV.12	The electronegativity difference between two bonded atoms is used to assess the degree of polarity in the bond.	distinguish between nonpolar covalent bonds (two of the same nonmetals) and polar covalent bonds	Bond polarity and ionic character increase with an increasing difference in electronegativity. Ionic character refers to an atom's tendency to transfer electrons when forming a chemical bond. Ionic character, like molecular polarity, is determined by finding the electronegativity difference between the atoms in the bond.
IV.13	Metals tend to react with nonmetals to form ionic compounds. Nonmetals tend to react with other nonmetals to form molecular (covalent) compounds. Ionic compounds containing polyatomic ions have both ionic and covalent bonding.		
IV.14	Ionic compounds will be soluble in water when the attraction between the ions and water molecules is greater than the attractions between ions in the solid compound.	predict solubility/insolubility of ionic compounds using Reference Table *F* identify precipitates in double replacement reactions	

V. Physical Behavior of Matter

	MAJOR UNDERSTANDINGS	**SKILLS** The student should be able to:	**KEY POINTS TO REMEMBER**
V.1	Matter is classified as a pure substance or as a mixture of substances.		A pure substance can either be an element or a compound.
V.2	The three phases of matter (solids, liquids, and gases) have different properties.	use a simple particle model to differentiate properties of a solid, a liquid, and a gas	

V. Physical Behavior of Matter (*Continued*)

		MAJOR UNDERSTANDINGS	SKILLS The student should be able to:	KEY POINTS TO REMEMBER
V.3		A pure substance (element or compound) has a constant composition and has constant properties throughout a given sample and from sample to sample.	use particle models/diagrams to differentiate elements, compounds, and mixtures	
V.4		Elements are substances that are composed of atoms that have the same atomic number. Elements cannot be broken down by chemical change.		
V.5		Mixtures are composed of two or more different substances that can be separated by physical means. When different substances are mixed together, a homogeneous or heterogeneous mixture is formed.		
V.6		The proportions of components in a mixture can be varied. Each component in a mixture retains its original properties.		
V.7		Differences in properties such as density, particle size, molecular polarity, boiling point and freezing point, and solubility permit physical separation of the components of the mixture.	describe the process and use of filtration, distillation, and chromatography in the separation of a mixture	
V.8		A solution is a homogeneous mixture of a solute dissolved in a solvent. The solubility of a solute in a given amount of solvent is dependent on the temperature, the pressure, and the chemical natures of the solute and solvent.	interpret and construct solubility curves use solubility curves to distinguish saturated, supersaturated, and unsaturated solutions apply the adage "like dissolves like" to real-world situations	Solid solutes increase in solubility and gas solutes decrease in solubility as the temperature increases.
V.9		The concentration of a solution may be expressed as: molarity (M), percent by volume, percent by mass, or parts per million (ppm).	describe the preparation of a solution, given the molarity interpret solution concentration data calculate solution concentrations in molarity (M), percent mass, and parts per million (ppm)	

V. Physical Behavior of Matter (*Continued*)

	MAJOR UNDERSTANDINGS	SKILLS The student should be able to:	KEY POINTS TO REMEMBER
V.10	The addition of a nonvolatile solute to a solvent causes the boiling point of the solvent to increase and the freezing point of the solvent to decrease. The greater the concentration of solute particles, the greater the effect.		Electrolytes have a greater effect on colligative properties than nonelectrolytes do because they dissociate into ions in aqueous solution.
V.11	Energy can exist in different forms, such as chemical, electrical, electromagnetic, heat, mechanical, and nuclear.		
V.12	Heat is a transfer of energy (usually thermal energy) from a body of higher temperature to a body of lower temperature. Thermal energy is associated with the random motion of atoms and molecules.	distinguish between heat energy and temperature in terms of molecular motion and amount of matter qualitatively interpret heating and cooling curves in terms of changes in kinetic and potential energy, heat of vaporization, heat of fusion, and phase changes	
V.13	Temperature is a measure of the average kinetic energy of the particles in a sample of matter. Temperature is not a form of energy.	distinguish between heat energy and temperature in terms of molecular motion and amount of matter explain phase changes in terms of the changes in energy and intermolecular distance convert between Kelvin and Celsius temperature scales using the formula on Reference Table *T*	
V.14	The concept of an ideal gas is a model to explain behavior of gases. A real gas is most like an ideal gas when the real gas is at low pressure and high temperature.		

V. Physical Behavior of Matter (*Continued*)

	MAJOR UNDERSTANDINGS	SKILLS The student should be able to:	KEY POINTS TO REMEMBER
V.15	Kinetic molecular theory (KMT) for an ideal gas states all gas particles: • are in random, constant, straight-line motion • are separated by great distances relative to their size; the volume of gas particles is considered negligible • have no attractive forces between them • have collisions that may result in a transfer of energy between particles, but the total energy of the system remains constant.		Pressure is caused by collisions of gas particles against the inner walls of a container.
V.16	Kinetic molecular theory describes the relationships of pressure, volume, temperature, velocity, and frequency and force of collisions among gas molecules.	explain the gas laws in terms of KMT solve problems, using the combined gas law	
V.17	Equal volumes of gases at the same temperature and pressure contain an equal number of particles.	convert temperatures in Celsius degrees (°C) to kelvins (K), and kelvins to Celsius degrees	The molar volume of a gas at STP is equal to 22.4 L/mol.
V.18	The concepts of kinetic and potential energy can be used to explain physical processes that include: fusion (melting); solidification (freezing); vaporization (boiling, evaporation), condensation, sublimation, and deposition.	qualitatively interpret heating and cooling curves in terms of changes in kinetic and potential energy, heat of vaporization, heat of fusion, and phase changes calculate the heat involved in a phase or temperature change for a given sample of matter explain phase change in terms of the changes in energy and intermolecular distances	At a substance's melting point, the solid and liquid phases are in equilibrium. At a substance's boiling point, the liquid and gas phases are in equilibrium.
V.19	A physical change results in the rearrangement of existing particles in a substance. A chemical change results in the formation of different substances with changed properties.		

V. Physical Behavior of Matter (*Continued*)

	MAJOR UNDERSTANDINGS	**SKILLS** The student should be able to:	**KEY POINTS TO REMEMBER**
V.20	Chemical and physical changes can be exothermic or endothermic.	distinguish between endothermic and exothermic reactions, using energy terms in a reaction equation, ΔH, potential energy diagrams or experimental data	Exothermic reactions have a $-\Delta H$. Endothermic reactions have a $+\Delta H$.
V.21	The structure and arrangement of particles and their interactions determine the physical state of a substance at a given temperature and pressure.	use a simple particle model to differentiate properties of solids, liquids, and gases	
V.22	Intermolecular forces created by the unequal distribution of change result in varying degrees of attraction between molecules. Hydrogen bonding is an example of a strong intermolecular force.	explain vapor pressure, evaporation rate, and phase changes in terms of intermolecular forces use vapor pressure curves (Reference Table *H*) to determine the boiling point of a substance	
V.23	Physical properties of substances can be explained in terms of chemical bonds and intermolecular forces. These properties include conductivity, malleability, solubility, hardness, melting point, and boiling point.	compare the physical properties of substances based upon chemical bonds and intermolecular forces	

VI. Kinetics/Equilibrium

	MAJOR UNDERSTANDINGS	**SKILLS** The student should be able to:	**KEY POINTS TO REMEMBER**
VI.1	Collision theory states that a reaction is most likely to occur if reactant particles collide with the proper energy and orientation.	use collision theory to explain how various factors, such as temperature, surface area, and concentration, influence the rate of reaction	The reaction rate depends upon the number of effective collisions between reacting particles.
VI.2	The rate of a chemical reaction depends on several factors: temperature, concentration, nature of reactants, surface area, and the presence of a catalyst.		Reactions between ionic substances are faster than reactions between molecular substances.
VI.3	Some chemical and physical changes can reach equilibrium.	identify examples of physical equilibria as solution equilibrium and phase equilibrium, including the concept that a saturated solution is at equilibrium	

VI. Kinetics/Equilibrium (*Continued*)

	MAJOR UNDERSTANDINGS	SKILLS The student should be able to:	KEY POINTS TO REMEMBER
VI.4	At equilibrium, the rate of the forward reaction equals the rate of the reverse reaction. The measurable quantities of reactants and products remain constant at equilibrium.	describe the concentration of particles and rates of opposing reactions in an equilibrium system	Equilibrium means equal and opposite rates, *not* equal amounts!
VI.5	Le Châtelier's principle can be used to predict the effect of stress (change in pressure, volume, concentration, and temperature) on a system at equilibrium.	qualitatively describe the effect of stress on equilibrium, using Le Châtelier's principle	A "shift to the right" means that product is being produced and reactants are being used up.
VI.6	Energy released or absorbed by a chemical reaction can be represented by a potential energy diagram.	read and interpret potential energy diagrams: PE of reactants and products, activation energy (with or without a catalyst), heat of reaction	
VI.7	Energy released or absorbed by a chemical reaction (heat of reaction) is equal to the difference between the potential energy of the products and the potential energy of the reactants.	use Reference Table *I* to find the heat of reaction for specified moles produced from a particular reaction	
VI.8	A catalyst provides an alternate reaction pathway that has a lower activation energy than an uncatalyzed reaction.	draw a line (to show the effect of adding a catalyst) on a potential energy diagram	
VI.9	Entropy is a measure of the randomness or disorder of a system. A system with greater disorder has greater entropy.	compare the entropy of phases of matter	
VI.10	Systems in nature tend to undergo changes toward lower energy and higher entropy.		

VII. Organic Chemistry

	MAJOR UNDERSTANDINGS	SKILLS The student should be able to:	KEY POINTS TO REMEMBER
VII.1	Organic compounds contain carbon atoms that bond to one another in chains, rings, and networks to form a variety of structures. Organic compounds can be named using the IUPAC system.	classify an organic compound based on its structural or condensed structural formula	

VII. Organic Chemistry (*Continued*)

	MAJOR UNDERSTANDINGS	SKILLS The student should be able to:	KEY POINTS TO REMEMBER
VII.2	Hydrocarbons are compounds that contain only carbon and hydrogen. Saturated hydrocarbons contain only single carbon–carbon bonds. Unsaturated hydrocarbons contain at least one multiple carbon–carbon bond.	draw structural formulas for alkanes, alkenes, and alkynes containing a maximum of ten carbon atoms use Reference Table Q to classify hydrocarbons as alkanes, alkenes, or alkynes	
VII.3	Organic acids, alcohols, esters, aldehydes, ketones, ethers, halides, amines, amides, and amino acids are types of organic compounds that differ in their structures. Functional groups impart distinctive physical and chemical properties to organic compounds.	classify an organic compound based on its structural or condensed structural formula draw a structural formula with the functional group(s) on a straight chain hydrocarbon backbone, when given the correct IUPAC name for the compound	Use Reference Table R for help in naming and drawing organic compounds.
VII.4	Isomers of organic compounds have the same molecular formula, but different structures and properties.		
VII.5	In a multiple covalent bond, more than one pair of electrons are shared between two atoms. Unsaturated organic compounds contain at least one double or triple bond.		
VII.6	Types of organic reactions include: addition, substitution, polymerization, esterification, fermentation, saponification, and combustion.	identify types of organic reactions determine a missing reactant or product in a balanced equation	

VIII. Oxidation–Reduction

	MAJOR UNDERSTANDINGS	SKILLS The student should be able to:	KEY POINTS TO REMEMBER
VIII.1	An oxidation–reduction (redox) reaction involves transfer of electrons (e^-).	determine a missing reactant or product in a balanced equation	
VIII.2	Reduction is the gain of electrons.		If an element's oxidation number is getting more negative (or less positive), it is reducing (gaining electrons).

VIII. Oxidation–Reduction (*Continued*)

	MAJOR UNDERSTANDINGS	SKILLS The student should be able to:	KEY POINTS TO REMEMBER
VIII.3	A half-reaction can be written to represent reduction.	write and balance half-reactions for oxidation and reduction of free elements and their monatomic ions	Reduction half-reactions show how many electrons are gained.
VIII.4	Oxidation is the loss of electrons.		If an element's oxidation number is getting less negative (or more positive), it is oxidizing (losing electrons).
VIII.5	A half-reaction can be written to represent oxidation.	write and balance half-reactions for the oxidation and reduction of free elements and their monatomic ions	Oxidation half-reactions show how many electrons are lost.
VIII.6	In a redox reaction, the number of electrons lost is equal to the number of electrons gained.		Electrons lost = electrons gained means that charge is conserved.
VIII.7	Oxidation numbers (states) can be assigned to atoms and ions. Changes in oxidation numbers indicate that oxidation and reduction have occurred.	identify which reactions are redox reactions	
VIII.8	An electrochemical cell can be either voltaic or electrolytic. In an electrochemical cell, oxidation occurs at the anode and reduction at the cathode.	compare and contrast voltaic and electrolytic cells	
VIII.9	A voltaic cell spontaneously converts chemical energy to electrical energy.	identify and label the parts of a voltaic cell (cathode, anode, salt bridge) and direction of electron flow, given the reaction equation use an activity series to determine whether a redox reaction is spontaneous	Remember: Ions flow through the salt bridge. Electrons flow through the wire.
VIII.10	An electrolytic cell requires electrical energy to produce chemical change. This process is known as electrolysis.	identify and label the parts of an electrolytic cell (anode, cathode) and direction of electron flow, given the reaction equation	

IX. Acids, Bases, and Salts

	MAJOR UNDERSTANDINGS	**SKILLS** The student should be able to:	**KEY POINTS TO REMEMBER**
IX.1	Behavior of many acids and bases can be explained by the Arrhenius theory. Arrhenius acids and bases are electrolytes.	given properties, identify substances as Arrhenius acids or Arrhenius bases	
IX.2	An electrolyte is a substance that, when dissolved in water, forms a solution capable of conducting an electric current. The ability of a solution to conduct an electric current depends on the concentration of ions.		The three categories of compounds that are electrolytes are acids, bases, and salts (ionic compounds).
IX.3	Arrhenius acids yield H^+ (hydrogen ion) as the only positive ion in aqueous solution. The hydrogen ion may also be written as H_3O^+, hydronium ion.		
IX.4	Arrhenius bases yield OH^- (hydroxide ion) as the only negative ion in an aqueous solution.		
IX.5	In the process of neutralization, an Arrhenius acid and an Arrhenius base react to form salt and water.	write simple neutralization reactions when given the reactants	$[H^+] = [OH^-]$ at the equivalence point.
IX.6	Titration is a laboratory process in which a volume of solution of known concentration is used to determine the concentration of another solution.	calculate the concentration or volume of a solution, using titration data	
IX.7	There are alternate acid–base theories. One such theory states that an acid is an H+ donor and a base is an H+ acceptor.	identify acids and bases in a reversible reaction using the alternate theory	
IX.8	The acidity and alkalinity of an aqueous solution can be measured by its pH value. The relative level of acidity or alkalinity of a solution can be shown by using indicators.	interpret changes in acid–base indicator color identify solutions as acid, base, or neutral based upon the pH	The pH of pure water is 7. Acids have pH values less than 7, and bases have pH values greater than 7.
IX.9	On the pH scale, each decrease of one unit of pH represents a tenfold increase in hydronium ion concentration.		

X. Nuclear Chemistry

	MAJOR UNDERSTANDINGS	SKILLS The student should be able to:	KEY POINTS TO REMEMBER
X.1	Stability of isotopes is based on the ratio of the neutrons and protons in its nucleus. Although most nuclei are stable, some are unstable and spontaneously decay emitting radiation.		
X.2	Each radioactive isotope has a specific mode and rate of decay (half-life).	calculate the initial amount, the fraction remaining, or the half-life of a radioactive isotope, given two of the three variables use Reference Table *N* to find the decay mode for a particular radioisotope	
X.3	A change in the nucleus of an atom that converts it from one element to another is called transmutation. This can occur naturally or can be induced by the bombardment of the nucleus of high-energy particles.		Natural transmutation is also called radioactive decay.
X.4	Spontaneous decay can involve the release of alpha particles, beta particles, positrons, and/or gamma radiation from the nucleus of an unstable isotope. These emissions differ in mass, charge, ionizing power, and penetrating power.	determine decay mode and write nuclear equations showing alpha and beta decay	Radioactive decay continues until a stable isotope is reached.
X.5	Nuclear reactions include natural and artificial transmutation, fission, and fusion.	compare and contrast fission and fusion reactions	
X.6	There are benefits and risks associated with fission and fusion reactions.		Fusion generates much more energy than fission. Fission produces far more dangerous radioactive waste than fusion.
X.7	Nuclear reactions can be represented by equations that include symbols that represent atomic nuclei (with the mass number and atomic number), subatomic particles (with mass number and charge), and/or emissions such as gamma radiation.	complete nuclear equations; predict missing particles from nuclear equations	

X. Nuclear Chemistry (*Continued*)

	MAJOR UNDERSTANDINGS	SKILLS The student should be able to:	KEY POINTS TO REMEMBER
X.8	Energy released in a nuclear reaction (fission or fusion) comes from the fractional amount of mass converted into energy. Nuclear changes convert matter into energy.		
X.9	Energy released during nuclear reactions is much greater than the energy released during chemical reactions.		
X.10	There are inherent risks associated with radioactivity and the use of radioactive isotopes. Risks can include biological exposure, long-term storage and disposal, and nuclear accidents.		
X.11	Radioactive isotopes have many beneficial uses. Radioactive isotopes are used in medicine and industrial chemistry, e.g., radioactive dating, tracing chemical and biological processes, industrial measurement, nuclear power, and detection and treatment of diseases.	identify specific uses of some common radioisotopes, such as: I-131 in diagnosing and treating thyroid disorders; C-14 to C-12 ratio in dating living organisms; U-238 to Pb-206 ratio in dating geological formations; Co-60 in treating cancer	

QUESTION INDEX

What follows is an index to the examination questions that are explained in this book. The questions are indexed according to the Sequence numbers given in the Topic Outline found on pages 20–38.

Some questions embrace more than one topic; these questions are marked with a dagger (†).

SEQUENCE	JUNE 2016	AUGUST 2016	JUNE 2017	AUGUST 2017	JUNE 2018	AUGUST 2018	JUNE 2019	AUGUST 2019
M. Math Skills								
M.1		10						
M.2		40			83†			
M.3	69	79	33		69, 80	45	33, 70	
M.4								75
M.5				50†	31†			
R. Reading Skills								
R.1			76	16, 49, 68	55	13†, 28, 39†, 58†, 68	33	18†, 53†	58†, 82
R.2					65	66		60†, 61	54†, 66
R.3									
I. Atomic Concepts								
I.1		48				4		1
I.2			1	1	1	2		
I.3	2						1	
I.4	1						2	
I.5			3	2	2	31	52	2
I.6	29, 84	32	4, 55	3	34, 36	3, 83	3	3
I.7			2			1		4, 52
I.8					9	55	4	5
I.9	72	33	31	31	35	32	32	32
I.10	73, 74	1	5		31†, 32		31	31
I.11		2			4	54	5	
I.12	31	39	56		3	53†	62	33
I.13	32	30	32	32	5			53

SEQUENCE	JUNE 2016	AUGUST 2016	JUNE 2017	AUGUST 2017	JUNE 2018	AUGUST 2018	JUNE 2019	AUGUST 2019
II. Periodic Table								
II.1	3	6				5		
II.2			6, 54	6, 82		53†		51
II.3	33	11, 12	66	61	6	6		7
II.4	12			5	39			
II.5	54			64	7		6	
II.6	4	34		33	8	7	7	
II.7	5	3		62		57	8†, 34	8
II.8		63	51	63		58	8†, 35	
II.9	36	31			11, 51	56		6
III. Moles/Stoichiometry								
III.1	6†, 35	4, 15†	34	34	67	8, 34	9, 66	9, 34
III.2	34		7		74		10	10
III.3	70	51	72	57		35	75	
III.4			8, 35	35		36†	11, 68	35
III.5	52, 60	67	36, 69	69, 72	37, 77	73	36, 69	78
III.6	75	5	74	56	63		53†, 67	68, 74
III.7	37, 59		70	68	55	51	37	67
III.8	38	56	9	75	38	37		
IV. Chemical Bonding								
IV.1						9, 13		
IV.2			52			10		
IV.3	66	36	11, 37	51, 58			77	11, 36
IV.4		7		74†				12
IV.5	39†	8†	38	52	10†, 14	11		
IV.6	68	60, 61	13	8	33	38	73	69
IV.7	41	37	58	12		12		13
IV.8	67	9	57	7			38	70
IV.9				49, 76			39	
IV.10	51	62	59	48		67		71
IV.11	9		60	9		15		14
IV.12	39†	8†, 38	12		41, 75	39	12	
IV.13					54			
IV.14	7		53	11	56		71	

SEQUENCE	JUNE 2016	AUGUST 2016	JUNE 2017	AUGUST 2017	JUNE 2018	AUGUST 2018	JUNE 2019	AUGUST 2019
V. Physical Behavior of Matter								
V.1								
V.2	11, 56	13, 14	21, 41, 67	4		62†	40	19, 54†
V.3	48			36	12			15, 37
V.4		15†		10			13	
V.5				13	40		15	16
V.6			20				14	
V.7	14	53	10	14	15	52		17
V.8	42, 76	52, 70	82	37		59, 60, 61	56	59
V.9		17, 71	81	38	42	40, 70	72	39
V.10	43	73		15		71	57	
V.11			77			14		
V.12		18, 55	14, 62	16		17, 62†, 64		
V.13	19	54	15	18		63	55, 60†	72
V.14	15		17	19		16	58	38
V.15				17	16		16	18
V.16	63, 64, 65	16, 50	18, 50, 71	39	64, 65	42	59	73
V.17	17, 55	59	19	21	43	18		40
V.18	58	41	63, 76	22	17, 60, 62	41	17, 41	56, 57
V.19		49		23†, 40, 41, 67		65		
V.20	71			50†, 66			18†	
V.21					61		19	
V.22	57†	84†			13, 52	19†		58†
V.23	8, 53, 57†	84†	39		10†	19†		55
VI. Kinetics/Equilibrium								
VI.1		19		20	19	43		41
VI.2	13	42, 69		70	72	68	42	
VI.3		20		23†			20	
VI.4	40	21	22	53	71	66	43	20
VI.5	44		42	54		24	44	42
VI.6			43			72, 74		76

SEQUENCE	JUNE 2016	AUGUST 2016	JUNE 2017	AUGUST 2017	JUNE 2018	AUGUST 2018	JUNE 2019	AUGUST 2019
VI.7		22	23	71	44	69	21	43
VI.8	18		24		73		22	77
VI.9	20		75	24				44
VI.10		23			20	20	23	21
VII. Organic Chemistry								
VII.1	21, 47	81	61	73	18	76	24	79
VII.2	45	83	73	59	21	22, 44, 77	25, 46	22, 80, 81
VII.3	23, 77	43, 57, 85	25, 44	42, 60, 77†	53, 70, 76	21, 75	45, 51, 76	23, 45
VII.4			26	77†	22			
VII.5	22			74†				
VII.6	50	82	27	43	45	23	47	
VIII. Oxidation–Reduction								
VIII.1	10				48		74	
VIII.2		24					26	
VIII.3	79	68	79		47		79†	62
VIII.4		58						
VIII.5						46		
VIII.6		35, 75	28	25	79	36†		24
VIII.7	24	74		80	46, 78	25	48	46
VIII.8		25		26		26	79†	25
VIII.9	78, 80	44, 66	78, 80	44	23, 49	47	78, 80†	47, 60, 61
VIII.10	25, 81		29	78	24		80†	
IX. Acids, Bases, and Salts								
IX.1			30†				27	
IX.2	49	26, 72	83	27, 79		48	85	26
IX.3	6†		30†			78		28
IX.4		78	30†	81†	25			
IX.5	26	45	40		26	79	83	48, 64
IX.6	46	80		45	82	80	84	63
IX.7	27	46	45	46	50		49	
IX.8	61	47, 77	84	81†	83†, 84	81	82	27
IX.9	62	27	85	47	81, 85	49	81	65

SEQUENCE	JUNE 2016	AUGUST 2016	JUNE 2017	AUGUST 2017	JUNE 2018	AUGUST 2018	JUNE 2019	AUGUST 2019
X. Nuclear Chemistry								
X.1		28			27	27		
X.2	28	29	46	84	58, 59	85	54, 65	84
X.3		64	65	85				29
X.4	82, 83	65	64		57	28, 84	63	83, 85
X.5	85		48	28		50, 82	28, 50	30
X.6								
X.7			47	83			64	
X.8	30			29	29	29		50
X.9							29	
X.10								49
X.11				30	30	30	30	

Glossary of Important Terms

Absolute zero The lowest possible temperature, written as 0 K or $-273°C$.

Accuracy The closeness of a measurement to an accepted value; see also **precision**.

Acid See **Arrhenius acid; Brønsted-Lowry acid**.

Activated complex The intermediate state between reactants and products in a chemical reaction; the peak of the potential energy diagram.

Activation energy The minimum energy needed to initiate a reaction.

Addition polymerization The joining of unsaturated monomers by a series of addition reactions.

Addition reaction The process in which a substance reacts across a double or triple bond in an organic compound.

Alcohol An organic compound containing a hydroxyl (—OH) group.

Aldehyde An organic compound containing a carbonyl group with at least one hydrogen atom attached to the carbonyl carbon.

Alkali metal Any Group 1 element, excluding hydrogen.

Alkaline earth element Any Group 2 element.

Alkane A hydrocarbon containing only single bonds between adjacent carbon atoms.

Alkene A hydrocarbon containing one double bond between two adjacent carbon atoms.

Alkyl group An open-chained hydrocarbon less one hydrogen atom; for example, CH_3 = methyl group, C_2H_5 = ethyl group. Unspecified alkyl groups are designed by the letter R.

Alkyne A hydrocarbon containing one triple bond between two adjacent carbon atoms.

Reprinted with permission from *Let's Review:* Chemistry by Albert S. Tarendash, © 1998 by Barron's Educational Series, Inc.

Allotrope A specific form of an element that can exist in more than one form; graphite and diamond are allotropes of the element carbon.

Alloy A solid metallic solution.

Alpha decay The radioactive process in which an alpha particle is emitted.

Alpha particle (α) A helium-4 nucleus.

Amide An organic compound containing the $CONH_2$ functional group.

Amine A hydrocarbon derivative containing an amino group.

Amino acid An organic compound containing at least one amino group and one carboxyl group.

Amino group An ammonia molecule less one hydrogen atom; —NH_2.

Anhydrous Pertaining to a compound from which the water of crystallization has been removed.

Anode The electrode at which oxidation occurs.

Aqueous Pertaining to a solution in which water is the solvent.

Aromatic hydrocarbon Any ring hydrocarbon whose electronic structure is related to that of benzene.

Arrhenius acid Any substance that releases H^+ ions in water.

Arrhenius base Any substance that releases OH^- ions in water.

Atmospheric pressure 1 standard atmosphere (atm) = 101.3 kilopascals.

Atom The basic unit of an element.

Atomic mass The weighted average of the masses of the isotopes of an element.

Atomic mass unit (u) One-twelfth the mass of a carbon-12 atom.

Atomic number The number of protons in the nucleus of an atom; the atomic number defines the element.

Atomic radius A measure of the size of an atom.

Avogadro's hypothesis Equal volumes of gases, measured at the same temperature and pressure, contain equal numbers of particles.

Avogadro's number (N_A) The number of particles in 1 mole; 6.02×10^{23}.

Battery A commercial voltaic cell.

Benzene C_6H_6; the parent hydrocarbon of all aromatic compounds.

Beta decay The radioactive process in which a beta particle is emitted.

Beta (−) particle (β^-) An electron.

Beta (+) particle (β^+) A positron.

Binary compound A compound containing two elements.

Binding energy The energy released when a nucleus is assembled from its nucleons.

Boiling The transition of liquid to gas; boiling occurs when the vapor pressure of a liquid equals the atmospheric pressure above the liquid.

Boiling point The temperature at which boiling occurs; the temperature at which the liquid and vapor phases of a substance are in equilibrium.

Boiling point elevation The increase in the boiling point of a solvent due to the presence of solute particles.

Bond energy The energy needed to break a chemical bond.

Boyle's law At constant temperature and mass, the pressure of an ideal gas is inversely proportional to its volume; $P_1 V_1 = P_2 V_2$.

Breeder reactor A fission reactor that generates its own nuclear fuel.

Bright-line spectrum The lines of visible light emitted by elements as electrons fall to lower energy levels.

Brønsted-Lowry acid A substance that can donate H^+ ions.

Brønsted-Lowry base A substance that can accept H^+ ions.

Carbonyl group The functional group characteristic of aldehydes and ketones; $> C=O$.

Carboxyl group The functional group characteristic of organic acids; —COOH.

Catalyst A substance that speeds a chemical reaction by lowering the activation energy of the reaction.

Cathode The electrode at which reduction occurs.

Celsius (C) scale The temperature scale on which the freezing and boiling points of water (at 1 atm) are set at 0 and 100, respectively.

Chain reaction A chemical or nuclear reaction in which one step supplies energy or reactants for the next step.

Charles's law At constant pressure and mass, the volume of an ideal gas is directly proportional to the Kelvin temperature; $\dfrac{V_1}{T_1} = \dfrac{V_2}{T_2}$.

Chemical bond The stabilizing of two atoms by sharing or transferring electrons.

Chemical equation A shorthand listing of reactants, products, and molar quantities in a chemical reaction.

Chemical equilibrium The state in which the rates of the forward and reverse reactions are equal.

Chemical family See **group**.

Coefficient A number in a chemical equation that indicates how many particles of a reactant or product are required or formed in the reaction.

Colligative property A property that depends on the number of particles present rather than the type of particle; see also **boiling point elevation; freezing point depression**.

Combined (ideal) gas law At constant mass, the product of the pressure and volume divided by the Kelvin temperature is a constant; $\dfrac{P_1 V_1}{T_1} = \dfrac{P_2 V_2}{T_2}$.

Compound A combination of two or more elements with a fixed composition by mass.

Concentrated Pertaining to a solution that contains a relatively large quantity of solute.

Concentration The "strength" of a solution; the quantity of solute relative to the quantity of solvent.

Condensation The change from gas to liquid.

Condensation polymerization The joining of monomers by a series of dehydration reactions.

Control rod The part of a fission reactor that controls the rate of fission by absorbing neutrons.

Coordinate covalent bond A single covalent bond in which the pair of electrons is supplied by one atom.

Covalent bond A chemical bond formed by the sharing of electrons.

Cracking The process of breaking large hydrocarbon molecules into smaller ones in order to increase the yield of compounds such as gasoline.

Crystal A solid whose particles are arranged in a regularly repeating pattern.

Decomposition A reaction in which a compound forms two or more simpler substances.

Density Mass per unit volume;

$$d = \frac{m}{V}.$$

Deposition The direct transition from gas to solid.

Deuterium The isotope of hydrogen with a mass number of 2.

Diatomic molecule A neutral particle consisting of two atoms; Br_2 and CO are diatomic molecules.

Diffusion The movement of one substance through another.

Dihydroxy alcohol An organic compound with two hydroxyl groups.

Dilute (adjective) Pertaining to a solution that contains a relatively small quantity of solute; (verb) to reduce the concentration of a solution by adding solvent.

Dipole An unsymmetrical charge distribution in a neutral molecule.

Dipole–dipole attraction The attractive force between two oppositely charged dipoles of neighboring polar molecules.

Dissociation The separation of an ionic compound in solution into positive and negative ions.

Distillation The simultaneous boiling of a liquid and condensation of its vapor.

Double bond A covalent bond in which two pairs of electrons are shared by two adjacent atoms.

Ductility The property of a substance that allows it to be drawn into a wire; metallic substances possess ductility.

Dynamic equilibrium The state in which the rates of opposing processes are equal; see also **chemical equilibrium; phase equilibrium; solution equilibrium**.

Electrochemical cell—either a *voltaic cell* or an electrolytic cell A device that produces usable electrical energy from a spontaneous redox reaction; see also **battery**.

Electrode A conductor in an electrochemical or electrolytic cell that serves as the site of oxidation or reduction.

Electrolysis A nonspontaneous redox reaction driven by an external source of electricity.

Electrolyte A substance whose aqueous solution conducts electricity.

Electrolytic cell A device for carrying out electrolysis.

Electron The elementary unit of negative charge.

Electron-dot diagram See **Lewis structure**.

Electronegativity The measure of an atom's attraction for a bonded pair of electrons.

Electroplating The use of an electric current to deposit a layer of metal on a negatively charged object.

Element A substance all of whose atoms have the same atomic number.

Empirical formula A formula in which the elements are present in the smallest whole-number ratio; NO_2 is an empirical formula, but C_2H_4 is not.

Endothermic reaction A reaction that absorbs energy; ΔH is positive for an endothermic reaction.

End point The point in a titration that signals that equivalent quantities of reactants have been added.

Energy A quantity related to an object's capacity to do work.

Enthalpy change (ΔH) The heat energy absorbed or released by a system at constant pressure.

Entropy (S) The measure of the randomness or disorder of a system.

Entropy change (ΔS) An increase or decrease in the randomness of a system.

Equilibrium See **dynamic equilibrium**.

Ester The organic product of esterification.

Esterification The reaction of an acid with an alcohol to produce an ester and water.

Ethanoic acid CH_3COOH; acetic acid.

Ethanol CH_3CH_2OH; ethyl (grain) alcohol.

Ethene C_2H_4; ethylene; the parent of the alkene family of hydrocarbons.

Ether An organic compound containing the arrangement R—O—R.

Ethyne C_2H_2; acetylene; the parent of the alkyne family of hydrocarbons.

Evaporation The surface transition of liquid to gas.

Excited state A condition in which one or more electrons in an atom are no longer in the lowest possible energy state.

Exothermic reaction A reaction that releases energy; ΔH is negative for an exothermic reaction.

Fermentation The (anaerobic) oxidation of a sugar such as glucose to produce ethanol and carbon dioxide; the reaction is catalyzed by enzymes.

Filtration A method of separating a liquid from the particles suspended in it.

First ionization energy The quantity of energy needed to remove the most loosely held electron from an isolated neutral atom.

Fission A nuclear reaction in which a heavy nuclide splits to form lighter nuclides and energy.

Fission reactor A device for producing electrical energy by means of a controlled fission reaction.

Formula mass The sum of the masses of the atoms in a formula; units are atomic mass units (u) or grams per mole (g/mol).

Fractional distillation The separation of organic substances based on differences in their boiling points.

Freezing The transition from liquid to solid.

Freezing point The temperature at which freezing occurs.

Freezing point depression (lowering) The decrease in the freezing point of a solvent due to the presence of solute particles.

Fuel rod The part of a nuclear reactor that contains the fissionable material.

Functional group An atom or group of atoms that confers specific properties on an organic molecule.

Fusion A synonym for *melting*; also, a nuclear process in which light nuclides join to form heavier nuclides and produce radiant energy.

Fusion reactor An experimental device for producing a controlled fusion reaction and generating electrical energy from it.

Gas The phase in which matter has neither definite shape nor definite volume.

Gram-atomic mass The molar mass of an element expressed in grams per mole (g/mol).

Gram-formula mass See **molar mass**.

Gram-molecular mass The molar mass of a molecule.

Ground state The electron configuration of an atom in the lowest energy state.

Group The elements within a single vertical column of the Periodic Table.

Half-cell The part of an electrochemical cell in which oxidation or reduction occurs.

Half-life The time needed for a substance to decay to one-half its initial mass.

Half-reaction The oxidation or reduction portion of a redox reaction.

Halogen An element in Group 17 of the Periodic Table; F, Cl, Br, I, At.

Heat energy The energy released or absorbed by a system undergoing a change in temperature, in phase, or in composition.

Heat of fusion (H_f) The heat energy absorbed when a unit mass of solid changes to liquid at its melting point; $H_{f(ice)} = 80$ calories per gram.

Heat of reaction (ΔH) The heat energy absorbed or released as a result of a chemical reaction.

Heat of vaporization (H_v) The heat energy absorbed when a unit mass of liquid changes to gas at its boiling point; $H_{v(water)} = 540$ calories per gram.

Heavy water A molecule of water in which the hydrogen atoms have a mass number of 2; deuterium oxide.

Heterogeneous mixture A nonuniform mixture.

Homogeneous mixture A mixture with a uniform distribution of particles; a solution is one example of a homogeneous mixture.

Homologous series A group of organic compounds with related structures and properties; each successive member of the series differs from the one before it by a specific number of carbon and hydrogen atoms (usually CH_2).

Hydrate A crystalline compound that has water molecules incorporated into its crystal structure; common examples include $CuSO_4 \cdot 5H_2O$ and $Na_2SO_4 \cdot 10H_2O$ [also written as $CuSO_4(H_2O)_5$ and $Na_2SO_4(H_2O)_{10}$].

Hydration The association of water molecules with an ion or another molecule.

Hydride A binary compound of an active metal and hydrogen; the oxidation state of hydrogen is -1.

Hydrocarbon An organic compound composed of carbon and hydrogen.

Hydrogen bond An unusually strong intermolecular attraction that results when hydrogen is bonded to a small, highly electronegative atom such as F, O, or N.

Hydrolysis A reaction in which a water molecule breaks a chemical bond; the reaction between certain salts and water to produce an excess of hydronium or hydroxide ions.

Hydronium ion H_3O^+; the conjugate acid of H_2O; responsible for acidic properties in water solutions.

Hydroxide ion OH^-; the conjugate base of H_2O; responsible for basic properties in water solutions.

Ideal gas A model of a gas in which the particles have no volume, do not attract or repel each other, and collide without loss of energy; real gases approximate ideal gas behavior under conditions of low pressure and high temperature.

Ideal gas law The relationship obeyed by an ideal gas; see **combined (ideal) gas law**.

Indicator A substance that undergoes a color change to signal a change in chemical conditions; acid–base indicators change color over specified pH ranges.

Inert (noble) gas An element in Group 18 of the Periodic Table; Ne, Ar, Kr, Xe, Rn. (He is also associated with Group 18.)

Inorganic compound A compound that is not a hydrocarbon derivative.

Ion A particle in which the numbers of protons and electrons are not equal.

Ion–dipole attraction The attractive force between an ion and the oppositely charged dipole of a neighboring polar molecule.

Ionic bond The electrostatic attraction of positive and negative ions in an ionic compound; an electronegativity difference of 1.7 or greater indicates the presence of an ionic bond.

Ionic compound A substance whose particles consist of positive and negative ions.

Ionization energy The quantity of energy needed to remove an electron from an atom or ion; see also **first ionization energy**.

Isomers Different compounds that have the same molecular formula.

Isotopes Atoms having the same atomic number but different mass numbers; atoms of the same element with differing numbers of neutrons.

IUPAC International Union of Pure and Applied Chemistry; the scientific group responsible for all major policies in chemistry, including the naming of elements and compounds.

Joule (J) The unit of work and energy in the SI (metric) system.

Kelvin (K) A measure of absolute temperature; the Kelvin scale begins at 0 and is related to the Celsius scale by the equation $K = C + 273$; a temperature *difference* of 1 K is equal to a temperature *difference* of 1°C.

Ketone An organic compound containing a carbonyl group with no hydrogen atoms directly attached to the carbonyl carbon.

Kilo- The metric prefix signifying 1,000.

Kilojoule (kJ) 1,000 joules.

Kinetic energy The energy associated with the motion of an object.

Kinetic molecular theory (KMT) The theory that explains the structure and behavior of idealized models of gases, liquids, and solids.

Le Châtelier's principle When a system at equilibrium is subjected to a stress, the system will shift in order to lessen the effects of the stress. Eventually, a new equilibrium point is established.

Lewis structure A shorthand notation for illustrating the ground-state valence electron configuration of an atom.

Liquid The phase in which matter has a definite volume but an indefinite shape; a liquid takes the shape of its container.

Liter (L) A unit of volume in the metric system; 1 liter = 1,000 cubic centimeters; 1 liter = 1 cubic decimeter; 1 liter is approximately equal to 1 quart.

Litmus An acid–base indicator that is red in acidic solutions and blue in basic solutions.

Macromolecule A giant molecule formed by network bonding or by polymerization.

Malleability The property by which a substance is able to be formed into various shapes; metallic substances possess malleability.

Mass number The number of nucleons in a nuclide.

Melting The transition from solid to liquid.

Melting point The temperature at which the vapor pressure of a solid equals the vapor pressure of the liquid; the temperature at which the solid and liquid phases of a substance are in equilibrium; see also **freezing point**.

Meniscus The curved surface of a liquid, caused by the attraction of the particles of the liquid and the container holding the liquid (e.g., water in a graduated cylinder), or by the mutual attraction of the particles of the liquid (e.g., mercury).

Metal A substance composed of atoms with low ionization energies and relatively vacant valence levels; Na, Fe, Ag, and Ba are metallic substances.

Metallic bond The delocalization of the valence electrons among the kernels of the metal atoms; "mobile valence electrons immersed in a sea of positive ions."

Metalloid An element that has both metallic and nonmetallic properties; examples of metalloids include B, Ge, Si, and Te.

Methanal HCHO; formaldehyde; the simplest aldehyde.

Methane CH_4; the parent of the alkane family of hydrocarbons.

Methanoic acid HCOOH; formic acid; the simplest organic acid.

Methanol CH_3OH; methyl (wood) alcohol; the simplest alcohol.

Milli- The metric prefix signifying 1/1,000.

Milliliter (mL) 1/1,000 liter; 1 liter = 1,000 milliliters.

Miscible A solution of liquids that is soluble in all proportions; ethanol and water are a miscible pair of liquids.

Mixture A material consisting of two or more components and having a variable composition.

Moderator A substance used to produce slow neutrons and promote nuclear fission; graphite and heavy water are used as moderators in fission reactors.

Molar mass The mass of any atom, element, ion, or compound expressed in grams per mole (g/mol).

Molarity The concentration of a solution, measured as the number of moles of solute per liter of solution.

Molar volume The volume occupied by 1 mole of an ideal gas; 22.4 liters at STP.

Mole The number of atoms contained in 12 grams of carbon-12; see also **Avogadro's number**.

Molecular formula A chemical formula that lists the number of atoms present but does not show the arrangement of the atoms in space.

Molecular mass The sum of the masses of the atoms in a molecule; units are atomic mass units (u) or grams per mole (g/mol).

Molecule The smallest unit of a nonionic substance; Ar, Cl_2, and NH_3 are molecules.

Monatomic molecule A molecule consisting of one atom; Xe and He are monatomic molecules.

Monomer The basic unit of a polymer; the monomer of a protein is an amino acid; the monomer of starch is glucose.

Network solid A substance formed by a two- or three-dimensional web of covalent bonds to produce a macromolecule; diamond and SiO_2 are network solids.

Neutralization The reaction of equivalent amounts of hydronium and hydroxide ions in aqueous solution; the principal product is water. When the water is evaporated, the spectator ions form a salt; see also **spectator ion**.

Neutron A neutral nuclear particle with a mass comparable to that of a proton.

Noble gas See **inert (noble) gas**.

Nonelectrolyte A substance whose aqueous solution does not conduct electricity; glucose is a nonelectrolyte.

Nonmetal A substance that does not have characteristic metallic properties; C and S are nonmetallic elements.

Nonpolar bond A covalent bond in which the electron pair or pairs are shared equally by both atoms.

Nonpolar molecule A molecule containing only nonpolar bonds, such as N_2, or a molecule with a symmetrical charge distribution, such as CCl_4 and CO_2.

Normal boiling point The boiling temperature of a substance at a pressure of 1 atmosphere.

Nuclear equation A shorthand listing of reactant and product nuclides in a nuclear reaction.

Nucleon A constituent of an atomic nucleus; a proton or a neutron.

Nucleus The portion of the atom that contains more than 99.9 percent of the atom's mass; the nucleus is small, dense, and positively charged.

Ore A native mineral from which a metal or metals can be extracted.

Organic chemistry The study of the hydrocarbons and their derivatives; the chemistry of carbon.

Organic compound A compound that is a hydrocarbon or a hydrocarbon derivative.

Oxidation The loss or apparent loss of electrons in a chemical reaction.

Oxidation number (state) The charge that an atom has or appears to have when certain arbitrary rules are applied; oxidation numbers are useful for identifying the atoms that are oxidized and reduced in a redox reaction.

Oxidizing agent The particle in a redox reaction that causes another particle to be oxidized; as a result, an oxidizing agent is reduced.

Paraffin A common name for a mixture of solid alkanes; another name for paraffin is wax.

Percent composition by mass The number of grams of an element (or group of elements) present in 100 grams of an ion or compound.

Period One of the horizontal rows of the Periodic Table; the period number indicates the valence level of an element.

Periodic law The properties of elements recur at regular intervals and depend on their nuclear charges; "The properties of elements are a periodic function of their atomic numbers."

Peroxide A compound in which oxygen has an oxidation state of -1; H_2O_2 and BaO_2 are peroxides.

Petroleum Crude oil containing a mixture of hydrocarbons.

pH A scale of acidity and basicity based on the hydronium ion concentration in aqueous solution; $pH = -\log[H_3O^+]$. At 298 K, a pH less than 7 indicates an acidic solution; a pH greater than 7, a basic solution; a pH of 7, a neutral solution.

Phase equilibrium The state in which the rates of opposing phase changes (freezing–melting, boiling–condensation, sublimation–deposition) are equal.

Phenolphthalein An acid–base indicator that is colorless in acidic solutions and pink in basic solutions.

Photon A fundamental unit of radiant energy; a quantum of radiation.

Polar bond A covalent bond in which the electron pair or pairs are shared unequally by both atoms; the atom with the larger electronegativity has more of the electron density surrounding it.

Polar molecule A molecule with an unsymmetrical charge distribution, such as H_2O; a dipole.

Polyatomic ion An ion composed of more than one atom; SO_4^{2-} is a polyatomic ion.

Polymer A macromolecule consisting of a chain of simpler units; polyethylene is a polymer of ethene.

Polymerization See **polymer**; **addition polymerization**; **condensation polymerization**.

Positron A positively charged electron; a particle of antimatter.

Potential energy The energy associated with the position of an object; a "stored" form of energy.

Precipitate A deposit formed by the appearance of an excess of solid solute in a saturated solution.

Precision The closeness of a series of measurements to one another; see also **accuracy**.

Pressure The force exerted on an object divided by the surface area of the object;

$$P = \frac{F}{A}.$$

Principal energy level An integer beginning with 1 that describes the approximate distance of an electron from the nucleus of an atom.

Product(s) The substance or substances that are formed in a chemical process; products are on the right side of a chemical equation.

Propanone $(CH_3)_2CO$; acetone; the simplest ketone.

Proton A nuclear particle with a positive charge equal to the negative charge on the electron; a nucleon.

Radiant energy Electromagnetic energy; visible light and X rays are examples of radiant energy.

Radioactive dating The use of radioactive isotopes to measure the age of an object.

Radioactive tracer A radioisotope used to indicate the path of an atom in a chemical reaction.

Radioactivity The spontaneous breakdown of a radioactive nuclide.

Radioisotope A radioactive isotope.

Reactant(s) The substance or substances that react in a chemical process; reactants are on the left side of a chemical equation.

Redox reaction A chemical reaction in which oxidation–reduction takes place.

Reducing agent The particle in a redox reaction that causes another particle to be reduced; as a result, a reducing agent is oxidized.

Reduction The gain or apparent gain of electrons in a chemical reaction.

Salt The spectator-ion product of a neutralization reaction; see also **spectator ion**.

Salt bridge A device for allowing the flow of ions in an electrochemical cell.

Saponification The reaction of an ester with a base to produce an alcohol and the sodium salt of an organic acid; soap is produced by saponifying fats with NaOH.

Saturated hydrocarbon A hydrocarbon containing only single carbon–carbon bonds.

Saturated solution A solution in which the pure solute is in equilibrium with the dissolved solute; a solution that contains the maximum amount of dissolved solute.

Significant digit(s) [figure(s)] The number or numbers that are part of a measurement. If there are two or more, all but the last figure is known; the last figure is the experimenter's best estimate.

Single bond A covalent bond in which one pair of electrons is shared by two adjacent atoms.

Solid The phase in which matter has both definite shape and definite volume.

Solubility The amount of solute needed to produce a saturated solution with a given amount of solvent.

Solute(s) The substance or substances dissolved in a solution.

Solution A homogeneous mixture whose particles are extremely small.

Solution equilibrium The state in which the undissolved and dissolved solutes are in dynamic equilibrium; in a solid–liquid solution, the rate of dissolving equals the rate of crystallization.

Solvent The substance in which the solute is dissolved.

Spectator ion An ion that does not take part in a chemical reaction; in the (acid–base) reaction between NaOH(aq) and HCl(aq), Na^+(aq) and Cl^-(aq) are spectator ions; in the (redox) reaction between Zn(s) and $Cu(NO_3)_2$(aq), NO_3^-(aq) is the spectator ion.

Standard solution A solution whose concentration is accurately known; a standard solution is used for analyzing other substances.

Standard temperature and pressure (STP) 273 K and 1 atmosphere.

Stock system A systematic method of naming chemical compounds in which the *positive* oxidation number is written as a Roman numeral in parentheses after the element. For example, the Stock name of the compound Fe_2O_3 is iron(III) oxide.

Stoichiometry The study of quantitative relationships in substances and reactions; chemical mathematics.

Strong acid In aqueous solution, a substance that ionizes almost completely to hydronium ion.

Strong base In aqueous solution, a substance that ionizes or dissociates almost completely to hydroxide ion.

Structural formula A chemical formula that illustrates the spatial arrangement of each atom.

Sublimation The direct transition from solid to gas; $CO_2(s)$ and $I_2(s)$ sublime at atmospheric pressure.

Substance An element or a compound.

Supersaturated solution A solution that contains more dissolved solute than a saturated solution at the same temperature; supersaturation is an unstable condition.

Temperature A measure of the average kinetic energy of the particles of a substance.

Titration The addition of a known volume of a standard solution in order to determine the concentration of an unknown solution.

Toluene C_7H_8; methylbenzene.

Transition element An element whose atoms contain unfilled d sublevels; an element in Groups 3–11 of the Periodic Table.

Transmutation The conversion of one element to another by a nuclear process.

Triple bond A covalent bond in which three pairs of electrons are shared by two adjacent atoms.

Tritium The radioactive isotope of hydrogen with a mass number of 3.

Unsaturated hydrocarbon A hydrocarbon containing double and/or triple carbon–carbon bonds.

Unsaturated solution A solution that contains less dissolved solute than a saturated solution at the same temperature.

Valence electron An electron in the outermost principle energy level.

Voltaic cell A device that produces usable electrical energy from a spontaneous redox reaction; see also **battery**.

Water of hydration The water molecules that are part of the crystalline structure of certain compounds.

Weak acid In aqueous solution, a substance that is poorly ionized and produces only a small concentration of hydronium ion.

Weak base In aqueous solution, a substance that is poorly ionized or dissociated and produces only a small concentration of hydroxide ion.

Reference Tables for Chemistry

Important Note: This is the 2011 Edition of the Reference Tables, which should be used for all Regents exams.

TABLE A
Standard Temperature and Pressure

Name	Value	Unit
Standard Pressure	101.3 kPa 1 atm	kilopascal atmosphere
Standard Temperature	273 K 0°C	kelvin degree Celsius

TABLE B
Physical Constants for Water

Heat of Fusion	334 J/g
Heat of Vaporization	2260 J/g
Specific Heat Capacity of $H_2O(\ell)$	4.18 J/g•K

TABLE C
Selected Prefixes

Factor	Prefix	Symbol
10^3	kilo-	k
10^{-1}	deci-	d
10^{-2}	centi-	c
10^{-3}	milli-	m
10^{-6}	micro-	μ
10^{-9}	nano-	n
10^{-12}	pico-	p

TABLE D
Selected Units

Symbol	Name	Quantity
m	meter	length
g	gram	mass
Pa	pascal	pressure
K	kelvin	temperature
mol	mole	amount of substance
J	joule	energy, work, quantity of heat
s	second	time
min	minute	time
h	hour	time
d	day	time
y	year	time
L	liter	volume
ppm	parts per million	concentration
M	molarity	solution concentration
u	atomic mass unit	atomic mass

TABLE E
Selected Polyatomic Ions

Formula	Name	Formula	Name
H_3O^+	hydronium	CrO_4^{2-}	chromate
Hg_2^{2+}	mercury(I)	$Cr_2O_7^{2-}$	dichromate
NH_4^+	ammonium	MnO_4^-	permanganate
$C_2H_3O_2^-$ CH_3COO^- } acetate		NO_2^-	nitrite
		NO_3^-	nitrate
CN^-	cyanide	O_2^{2-}	peroxide
CO_3^{2-}	carbonate	OH^-	hydroxide
HCO_3^-	hydrogen carbonate	PO_4^{3-}	phosphate
$C_2O_4^{2-}$	oxalate	SCN^-	thiocyanate
ClO^-	hypochlorite	SO_3^{2-}	sulfite
ClO_2^-	chlorite	SO_4^{2-}	sulfate
ClO_3^-	chlorate	HSO_4^-	hydrogen sulfate
ClO_4^-	perchlorate	$S_2O_3^{2-}$	thiosulfate

TABLE F
Solubility Guidelines for Aqueous Solutions

Ions That Form *Soluble* Compounds	Exceptions
Group 1 ions (Li^+, Na^+, etc.)	
ammonium (NH_4^+)	
nitrate (NO_3^-)	
acetate ($C_2H_3O_2^-$ or CH_3COO^-)	
hydrogen carbonate (HCO_3^-)	
chlorate (ClO_3^-)	
halides (Cl^-, Br^-, I^-)	when combined with Ag^+, Pb^{2+}, or Hg_2^{2+}
sulfates (SO_4^{2-})	when combined with Ag^+, Ca^{2+}, Sr^{2+}, Ba^{2+}, or Pb^{2+}

Ions That Form *Insoluble* Compounds*	Exceptions
carbonate (CO_3^{2-})	when combined with Group 1 ions or ammonium (NH_4^+)
chromate (CrO_4^{2-})	when combined with Group 1 ions, Ca^{2+}, Mg^{2+}, or ammonium (NH_4^+)
phosphate (PO_4^{3-})	when combined with Group 1 ions or ammonium (NH_4^+)
sulfide (S^{2-})	when combined with Group 1 ions or ammonium (NH_4^+)
hydroxide (OH^-)	when combined with Group 1 ions, Ca^{2+}, Ba^{2+}, Sr^{2+}, or ammonium (NH_4^+)

*compounds having very low solubility in H_2O

TABLE G
Solubility Curves at Standard Pressure

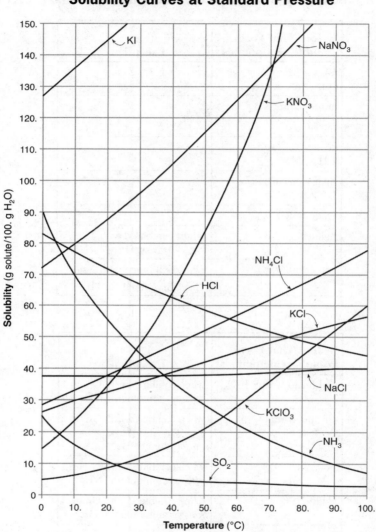

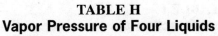

TABLE H
Vapor Pressure of Four Liquids

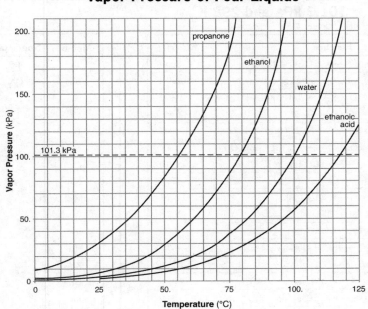

TABLE I
Heats of Reaction at
101.3 kPa and 298 K

Reaction	ΔH (kJ)*
$CH_4(g) + 2O_2(g) \longrightarrow CO_2(g) + 2H_2O(\ell)$	−890.4
$C_3H_8(g) + 5O_2(g) \longrightarrow 3CO_2(g) + 4H_2O(\ell)$	−2219.2
$2C_8H_{18}(\ell) + 25O_2(g) \longrightarrow 16CO_2(g) + 18H_2O(\ell)$	−10943
$2CH_3OH(\ell) + 3O_2(g) \longrightarrow 2CO_2(g) + 4H_2O(\ell)$	−1452
$C_2H_5OH(\ell) + 3O_2(g) \longrightarrow 2CO_2(g) + 3H_2O(\ell)$	−1367
$C_6H_{12}O_6(s) + 6O_2(g) \longrightarrow 6CO_2(g) + 6H_2O(\ell)$	−2804
$2CO(g) + O_2(g) \longrightarrow 2CO_2(g)$	−566.0
$C(s) + O_2(g) \longrightarrow CO_2(g)$	−393.5
$4Al(s) + 3O_2(g) \longrightarrow 2Al_2O_3(s)$	−3351
$N_2(g) + O_2(g) \longrightarrow 2NO(g)$	+182.6
$N_2(g) + 2O_2(g) \longrightarrow 2NO_2(g)$	+66.4
$2H_2(g) + O_2(g) \longrightarrow 2H_2O(g)$	−483.6
$2H_2(g) + O_2(g) \longrightarrow 2H_2O(\ell)$	−571.6
$N_2(g) + 3H_2(g) \longrightarrow 2NH_3(g)$	−91.8
$2C(s) + 3H_2(g) \longrightarrow C_2H_6(g)$	−84.0
$2C(s) + 2H_2(g) \longrightarrow C_2H_4(g)$	+52.4
$2C(s) + H_2(g) \longrightarrow C_2H_2(g)$	+227.4
$H_2(g) + I_2(g) \longrightarrow 2HI(g)$	+53.0
$KNO_3(s) \xrightarrow{H_2O} K^+(aq) + NO_3^-(aq)$	+34.89
$NaOH(s) \xrightarrow{H_2O} Na^+(aq) + OH^-(aq)$	−44.51
$NH_4Cl(s) \xrightarrow{H_2O} NH_4^+(aq) + Cl^-(aq)$	+14.78
$NH_4NO_3(s) \xrightarrow{H_2O} NH_4^+(aq) + NO_3^-(aq)$	+25.69
$NaCl(s) \xrightarrow{H_2O} Na^+(aq) + Cl^-(aq)$	+3.88
$LiBr(s) \xrightarrow{H_2O} Li^+(aq) + Br^-(aq)$	−48.83
$H^+(aq) + OH^-(aq) \longrightarrow H_2O(\ell)$	−55.8

*The ΔH values are based on molar quantities represented in the equations. A minus sign indicates an exothermic reaction.

TABLE J
Activity Series**

Most Active	Metals	Nonmetals	Most Active
	Li	F_2	
	Rb	Cl_2	
	K	Br_2	
	Cs	I_2	
	Ba		
	Sr		
	Ca		
	Na		
	Mg		
	Al		
	Ti		
	Mn		
	Zn		
	Cr		
	Fe		
	Co		
	Ni		
	Sn		
	Pb		
	H_2		
	Cu		
	Ag		
Least Active	Au		Least Active

**Activity Series is based on the hydrogen standard. H_2 is *not* a metal.

TABLE K
Common Acids

Formula	Name
$HCl(aq)$	hydrochloric acid
$HNO_2(aq)$	nitrous acid
$HNO_3(aq)$	nitric acid
$H_2SO_3(aq)$	sulfurous acid
$H_2SO_4(aq)$	sulfuric acid
$H_3PO_4(aq)$	phosphoric acid
$H_2CO_3(aq)$ or $CO_2(aq)$	carbonic acid
$CH_3COOH(aq)$ or $HC_2H_3O_2(aq)$	ethanoic acid (acetic acid)

TABLE L
Common Bases

Formula	Name
$NaOH(aq)$	sodium hydroxide
$KOH(aq)$	potassium hydroxide
$Ca(OH)_2(aq)$	calcium hydroxide
$NH_3(aq)$	aqueous ammonia

TABLE M
Common Acid–Base Indicators

Indicator	Approximate pH Range for Color Change	Color Change
methyl orange	3.1–4.4	red to yellow
bromthymol blue	6.0–7.6	yellow to blue
phenolphthalein	8–9	colorless to pink
litmus	4.5–8.3	red to blue
bromcresol green	3.8–5.4	yellow to blue
thymol blue	8.0–9.6	yellow to blue

Source: *The Merck Index*, 14th ed., 2006, Merck Publishing Group

TABLE N
Selected Radioisotopes

Nuclide	Half-Life	Decay Mode	Nuclide Name
^{198}Au	2.695 d	β^-	gold-198
^{14}C	5715 y	β^-	carbon-14
^{37}Ca	182 ms	β^+	calcium-37
^{60}Co	5.271 y	β^-	cobalt-60
^{137}Cs	30.2 y	β^-	cesium-137
^{53}Fe	8.51 min	β^+	iron-53
^{220}Fr	27.4 s	α	francium-220
^{3}H	12.31 y	β^-	hydrogen-3
^{131}I	8.021 d	β^-	iodine-131
^{37}K	1.23 s	β^+	potassium-37
^{42}K	12.36 h	β^-	potassium-42
^{85}Kr	10.73 y	β^-	krypton-85
^{16}N	7.13 s	β^-	nitrogen-16
^{19}Ne	17.22 s	β^+	neon-19
^{32}P	14.28 d	β^-	phosphorus-32
^{239}Pu	2.410×10^4 y	α	plutonium-239
^{226}Ra	1599 y	α	radium-226
^{222}Rn	3.823 d	α	radon-222
^{90}Sr	29.1 y	β^-	strontium-90
^{99}Tc	2.13×10^5 y	β^-	technetium-99
^{232}Th	1.40×10^{10} y	α	thorium-232
^{233}U	1.592×10^5 y	α	uranium-233
^{235}U	7.04×10^8 y	α	uranium-235
^{238}U	4.47×10^9 y	α	uranium-238

Source: *CRC Handbook of Chemistry and Physics*, 91st ed., 2010–2011, CRC Press

TABLE O
Symbols Used in Nuclear Chemistry

Name	Notation	Symbol
alpha particle	^4_2He or $^4_2\alpha$	α
beta particle	$^0_{-1}e$ or $^0_{-1}\beta$	β^-
gamma radiation	$^0_0\gamma$	γ
neutron	^1_0n	n
proton	^1_1H or ^1_1p	p
positron	$^0_{+1}e$ or $^0_{+1}\beta$	β^+

TABLE P
Organic Prefixes

Prefix	Number of Carbon Atoms
meth-	1
eth-	2
prop-	3
but-	4
pent-	5
hex-	6
hept-	7
oct-	8
non-	9
dec-	10

TABLE Q
Homologous Series of Hydrocarbons

Name	General Formula	Examples	
		Name	Structural Formula
alkanes	C_nH_{2n+2}	ethane	H H │ │ H─C─C─H │ │ H H
alkenes	C_nH_{2n}	ethene	H⠀⠀⠀⠀H ⠀╲⠀⠀╱ ⠀⠀C=C ⠀╱⠀⠀╲ H⠀⠀⠀⠀H
alkynes	C_nH_{2n-2}	ethyne	H─C≡C─H

Note: n = number of carbon atoms

TABLE R
Organic Functional Groups

Class of Compound	Functional Group	General Formula	Example
halide (halocarbon)	—F (fluoro-) —Cl (chloro-) —Br (bromo-) —I (iodo-)	$R-X$ (X represents any halogen)	$CH_3CHClCH_3$ 2-chloropropane
alcohol	—OH	$R-OH$	$CH_3CH_2CH_2OH$ 1-propanol
ether	—O—	$R-O-R'$	$CH_3OCH_2CH_3$ methyl ethyl ether
aldehyde	$\overset{\displaystyle O}{\overset{\displaystyle \|}{-C}}-H$	$R-\overset{\displaystyle O}{\overset{\displaystyle \|}{C}}-H$	$CH_3CH_2\overset{\displaystyle O}{\overset{\displaystyle \|}{C}}-H$ propanal
ketone	$-\overset{\displaystyle O}{\overset{\displaystyle \|}{C}}-$	$R-\overset{\displaystyle O}{\overset{\displaystyle \|}{C}}-R'$	$CH_3\overset{\displaystyle O}{\overset{\displaystyle \|}{C}}CH_2CH_2CH_3$ 2-pentanone
organic acid	$-\overset{\displaystyle O}{\overset{\displaystyle \|}{C}}-OH$	$R-\overset{\displaystyle O}{\overset{\displaystyle \|}{C}}-OH$	$CH_3CH_2\overset{\displaystyle O}{\overset{\displaystyle \|}{C}}-OH$ propanoic acid
ester	$-\overset{\displaystyle O}{\overset{\displaystyle \|}{C}}-O-$	$R-\overset{\displaystyle O}{\overset{\displaystyle \|}{C}}-O-R'$	$CH_3CH_2COCH_3$ methyl propanoate
amine	$-\overset{\displaystyle |}{N}-$	$R-\overset{\displaystyle R'}{\overset{\displaystyle |}{N}}-R''$	$CH_3CH_2CH_2NH_2$ 1-propanamine
amide	$-\overset{\displaystyle O}{\overset{\displaystyle \|}{C}}-\overset{\displaystyle |}{N}H$	$R-\overset{\displaystyle O}{\overset{\displaystyle \|}{C}}-\overset{\displaystyle R'}{\overset{\displaystyle |}{N}}H$	$CH_3CH_2\overset{\displaystyle O}{\overset{\displaystyle \|}{C}}-NH_2$ propanamide

Note: R represents a bonded atom or group of atoms.

Periodic Table of the Elements

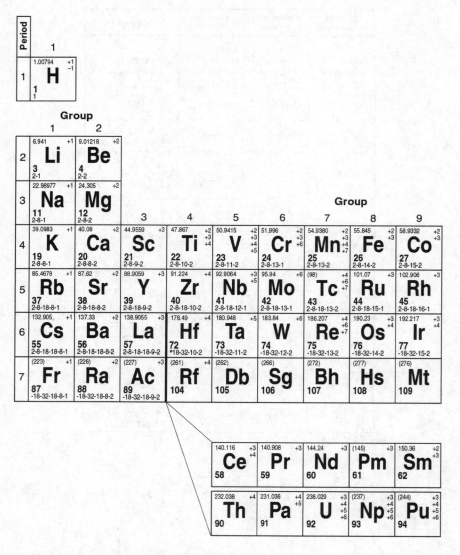

*denotes the presence of (2-8-) for elements 72 and above

**The systematic names and symbols for elements of atomic numbers 113 and above will be used until the approval of trivial names by IUPAC.

Source: *CRC Handbook of Chemistry and Physics*, 91st ed., 2010–2011, CRC Press

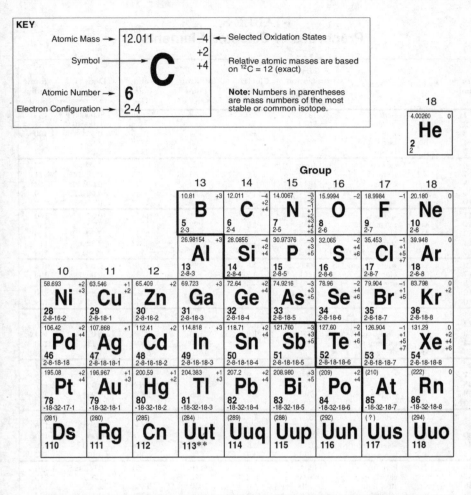

KEY

Atomic Mass → 12.011 −4 ← Selected Oxidation States

Symbol → **C** +2
  +4

Relative atomic masses are based on $^{12}C = 12$ (exact)

Atomic Number → **6**

Electron Configuration → 2-4

Note: Numbers in parentheses are mass numbers of the most stable or common isotope.

TABLE S
Properties of Selected Elements

Atomic Number	Symbol	Name	First Ionization Energy (kJ/mol)	Electro-negativity	Melting Point (K)	Boiling* Point (K)	Density** (g/cm³)	Atomic Radius (pm)
1	H	hydrogen	1312	2.2	14	20.	0.000082	32
2	He	helium	2372	—	—	4	0.000164	37
3	Li	lithium	520.	1.0	454	1615	0.534	130.
4	Be	beryllium	900.	1.6	1560.	2744	1.85	99
5	B	boron	801	2.0	2348	4273	2.34	84
6	C	carbon	1086	2.6	—	—	—	75
7	N	nitrogen	1402	3.0	63	77	0.001145	71
8	O	oxygen	1314	3.4	54	90.	0.001308	64
9	F	fluorine	1681	4.0	53	85	0.001553	60.
10	Ne	neon	2081	—	24	27	0.000825	62
11	Na	sodium	496	0.9	371	1156	0.97	160.
12	Mg	magnesium	738	1.3	923	1363	1.74	140.
13	Al	aluminum	578	1.6	933	2792	2.70	124
14	Si	silicon	787	1.9	1687	3538	2.3296	114
15	P	phosphorus (white)	1012	2.2	317	554	1.823	109
16	S	sulfur (monoclinic)	1000.	2.6	388	718	2.00	104
17	Cl	chlorine	1251	3.2	172	239	0.002898	100.
18	Ar	argon	1521	—	84	87	0.001633	101
19	K	potassium	419	0.8	337	1032	0.89	200.
20	Ca	calcium	590.	1.0	1115	1757	1.54	174
21	Sc	scandium	633	1.4	1814	3109	2.99	159
22	Ti	titanium	659	1.5	1941	3560.	4.506	148
23	V	vanadium	651	1.6	2183	3680.	6.0	144
24	Cr	chromium	653	1.7	2180.	2944	7.15	130.
25	Mn	manganese	717	1.6	1519	2334	7.3	129
26	Fe	iron	762	1.8	1811	3134	7.87	124
27	Co	cobalt	760.	1.9	1768	3200.	8.86	118
28	Ni	nickel	737	1.9	1728	3186	8.90	117
29	Cu	copper	745	1.9	1358	2835	8.96	122
30	Zn	zinc	906	1.7	693	1180.	7.134	120.
31	Ga	gallium	579	1.8	303	2477	5.91	123
32	Ge	germanium	762	2.0	1211	3106	5.3234	120.
33	As	arsenic (gray)	944	2.2	1090.	—	5.75	120.
34	Se	selenium (gray)	941	2.6	494	958	4.809	118
35	Br	bromine	1140.	3.0	266	332	3.1028	117
36	Kr	krypton	1351	—	116	120.	0.003425	116
37	Rb	rubidium	403	0.8	312	961	1.53	215
38	Sr	strontium	549	1.0	1050.	1655	2.64	190.
39	Y	yttrium	600.	1.2	1795	3618	4.47	176
40	Zr	zirconium	640.	1.3	2128	4682	6.52	164

TABLE S
Properties of Selected Elements (*Continued*)

Atomic Number	Symbol	Name	First Ionization Energy (kJ/mol)	Electro-negativity	Melting Point (K)	Boiling* Point (K)	Density** (g/cm³)	Atomic Radius (pm)
41	Nb	niobium	652	1.6	2750.	5017	8.57	156
42	Mo	molybdenum	684	2.2	2896	4912	10.2	146
43	Tc	technetium	702	2.1	2430.	4538	11	138
44	Ru	ruthenium	710.	2.2	2606	4423	12.1	136
45	Rh	rhodium	720.	2.3	2237	3968	12.4	134
46	Pd	palladium	804	2.2	1828	3236	12.0	130.
47	Ag	silver	731	1.9	1235	2435	10.5	136
48	Cd	cadmium	868	1.7	594	1040.	8.69	140.
49	In	indium	558	1.8	430.	2345	7.31	142
50	Sn	tin (white)	709	2.0	505	2875	7.287	140.
51	Sb	antimony (gray)	831	2.1	904	1860.	6.68	140.
52	Te	tellurium	869	2.1	723	1261	6.232	137
53	I	iodine	1008	2.7	387	457	4.933	136
54	Xe	xenon	1170.	2.6	161	165	0.005366	136
55	Cs	cesium	376	0.8	302	944	1.873	238
56	Ba	barium	503	0.9	1000.	2170.	3.62	206
57	La	lanthanum	538	1.1	1193	3737	6.15	194
				Elements 58–71 have been omitted.				
72	Hf	hafnium	659	1.3	2506	4876	13.3	164
73	Ta	tantalum	728	1.5	3290.	5731	16.4	158
74	W	tungsten	759	1.7	3695	5828	19.3	150.
75	Re	rhenium	756	1.9	3458	5869	20.8	141
76	Os	osmium	814	2.2	3306	5285	22.587	136
77	Ir	iridium	865	2.2	2719	4701	22.562	132
78	Pt	platinum	864	2.2	2041	4098	21.5	130.
79	Au	gold	890.	2.4	1337	3129	19.3	130.
80	Hg	mercury	1007	1.9	234	630.	13.5336	132
81	Tl	thallium	589	1.8	577	1746	11.8	144
82	Pb	lead	716	1.8	600.	2022	11.3	145
83	Bi	bismuth	703	1.9	544	1837	9.79	150.
84	Po	polonium	812	2.0	527	1235	9.20	142
85	At	astatine	—	2.2	575	—	—	148
86	Rn	radon	1037	—	202	211	0.009074	146
87	Fr	francium	393	0.7	300.	—	—	242
88	Ra	radium	509	0.9	969	—	5	211
89	Ac	actinium	499	1.1	1323	3471	10.	201
				Elements 90 and above have been omitted.				

* boiling point at standard pressure
** density of solids and liquids at room temperature and density of gases at 298 K and 101.3 kPa
— no data available
Source: *CRC Handbook for Chemistry and Physics*, 91ˢᵗ ed., 2010–2011, CRC Press

TABLE T
Important Formulas and Equations

Density	$d = \dfrac{m}{V}$	d = density m = mass V = volume
Mole Calculations	number of moles = $\dfrac{\text{given mass}}{\text{gram-formula mass}}$	
Percent Error	% error = $\dfrac{\text{measured value} - \text{accepted value}}{\text{accepted value}} \times 100$	
Percent Composition	% composition by mass = $\dfrac{\text{mass of part}}{\text{mass of whole}} \times 100$	
Concentration	parts per million = $\dfrac{\text{mass of solute}}{\text{mass of solution}} \times 1\,000\,000$	
	molarity = $\dfrac{\text{moles of solute}}{\text{liter of solution}}$	
Combined Gas Law	$\dfrac{P_1 V_1}{T_1} = \dfrac{P_2 V_2}{T_2}$	P = pressure V = volume T = temperature
Titration	$M_A V_A = M_B V_B$	M_A = molarity of H^+ M_B = molarity of OH^- V_A = volume of acid V_B = volume of base
Heat	$q = mC\Delta T$ $q = mH_f$ $q = mH_v$	q = heat H_f = heat of fusion m = mass H_v = heat of vaporization C = specific heat capacity ΔT = change in temperature
Temperature	$K = °C + 273$	K = kelvin $°C$ = degree Celsius

Using the Equations to Solve Chemistry Problems

IMPORTANT FORMULAS/EQUATIONS

In this section, various problems are solved using the formulas and equations found in Reference Table *T*.

1. Density: $d = \dfrac{m}{v}$

Problem

Calculate the mass of a sample of the element niobium (Atomic Number = 41) if the volume of the sample is 13.00 cm³.

Solution

- Use Reference Table *S* to find the density of niobium (8.570 g/cm³).
- Rearrange the variables in the equation in order to solve for mass: $m = d \cdot v$.
- Substitute the data in the equation and perform the calculation:

$$m = (8.570 \ \frac{\text{g}}{\text{cm}^3}) \cdot (13.00 \ \text{cm}^3) = \textbf{111.4 g}$$

2. Mole Calculations: number of moles $= \dfrac{\text{given mass}}{\text{gram-formula mass}}$

Problem

Calculate the number of moles in a sample of oxygen gas (O_2, gram-formula mass $= 32.0$ g/mol) if its mass is 56.0 grams.

Solution

- Substitute the data into the equation:

$$\text{number of moles} = \frac{56.0 \text{ g}}{32.0 \dfrac{\text{g}}{\text{mol}}} = \textbf{1.75 mol}$$

----

3. Percent Error: % error $= \dfrac{\text{measured value} - \text{accepted value}}{\text{accepted value}} \times 100$

Problem

A student measured the density of a sample of iron and obtained a value of 8.391 g/cm^3. What is the percent error of the student's determination?

Solution

- Use Reference Table S to obtain the density of iron (7.874 g/cm^3).
- Substitute the data into the equation:

$$\% \text{ error} = \frac{8.391 \dfrac{\text{g}}{\text{cm}^3} - 7.874 \dfrac{\text{g}}{\text{cm}^3}}{7.874 \dfrac{\text{g}}{\text{cm}^3}} \times 100 = \textbf{6.566\%}$$

----

4. Percent Composition: $\%\text{ composition by mass} = \dfrac{\text{mass of part}}{\text{mass of whole}} \times 100$

Problem

Calculate the percent composition by mass of carbon in $C_6H_{12}O_6$.

Solution

- Use the Periodic Table to determine the mass of one mole of the compound (180.2 g) and the mass of the *carbon* in 1 mole of the compound (72.07 g).
- Substitute the data into the equation:

$$\%\text{ composition} = \frac{72.07\,\text{g}}{180.2\,\text{g}} \times 100 = \textbf{39.99\%}$$

5a. Parts per Million: $\dfrac{\text{grams of solute}}{\text{grams of solution}} \times 1{,}000{,}000$

Problem

The concentration of arsenic in a certain river is 0.267 gram per 2,000 grams of river water. What is the arsenic concentration in parts per million (ppm)?

Solution

- Substitute the data into the equation:

$$\frac{0.267\,\text{g}}{2{,}000\,\text{g}} \times 1{,}000{,}000 = \textbf{134 ppm}$$

5b. Molarity: $\text{molarity} = \dfrac{\text{grams of solute}}{\text{liters of solution}}$

Problem

What is the molarity of an aqueous solution of ammonia (gram-formula mass = 17.03 g/mol) if 30.8 grams of solute are dissolved in 2.50 liters of solution?

Solution

- Use the gram-formula mass to convert the mass to moles.
- Substitute the data into the equation:

$$\text{molarity} = \frac{(30.8 \text{ g}) \cdot \left(\dfrac{1 \text{ mol}}{17.03 \text{ g}}\right)}{2.50 \text{ L}} = \boldsymbol{0.723 \text{ } M}$$

---

6. Combined Gas Law: $\dfrac{P_1 V_1}{T_1} = \dfrac{P_2 V_2}{T_2}$

Problem

A 50.0-milliliter sample, initially at STP, has its pressure changed to 0.85 atmosphere and its temperature changed to 330 K. What is the new volume of the gas?

Solution

- Use Reference Table A to obtain the STP values for temperature (273 K) and pressure (1 atm).
- Rearrange the equation in order to solve for V_2: $V_2 = V_2 = \dfrac{P_1 V_1 T_2}{T_1 P_2}$
- Substitute the data into the equation:

$$V_2 = \frac{(1 \text{ atm}) \cdot (50.0 \text{ mL}) \cdot (330. \text{ K})}{(273 \text{ K}) \cdot (0.85 \text{ atm})} = \boldsymbol{71.1 \text{ mL}}$$

---

7. Titration (Acid–Base): $M_A V_A = M_B V_B$

Problem

How many milliliters of 0.25-molar NaOH are needed to neutralize 75 milliliters of 0.17-molar HCl?

Solution

- Rearrange the equation to solve for V_B: $V_B = \dfrac{M_A M_A}{M_B}$.
- Substitute the data into the equation:

$$V_B = \frac{(0.17 \text{ M}) \cdot (75 \text{ mL})}{(0.25 \text{ M})} = \textbf{51 mL (NaOH)}$$

8a. Heat Transferred: $q = mC\Delta T$

Problem

How much heat is released by 120 grams of liquid water as it cools from 340 K to 290 K?

Solution

- Use Reference Table B to obtain the specific heat capacity of liquid water (4.2 J/g·K).
- Substitute the data into the equation:

$$q = (120 \text{ g}) \cdot \left(4.2 \ \frac{\text{J}}{\text{g·K}}\right) \cdot (290 \text{ K} - 340 \text{ K}) = \textbf{−25,000 J}$$

(The minus sign indicates that heat is *released*.)

8b. Heat of Fusion: $q = mH_f$

Problem

How much heat is absorbed by 200.0 grams of ice as it melts at 273 K?

Solution

- Use Reference Table *B* to obtain the value for the heat of fusion of ice (333.6 J/g).
- Substitute the data into the equation:

$$q = (200.0 \text{ g}) \cdot \left(333.6 \ \frac{\text{J}}{\text{g}}\right) = \mathbf{66,720 \ J}$$

8c. Heat of Vaporization: $q = mH_v$

Problem

How many grams of steam at 100°C will be condensed to liquid water if 4,000. joules of heat are released?

Solution

- Use Reference Table *B* to obtain the value for the heat of vaporization of water (2,259 J/g).
- Rearrange the equation to solve for mass: $m = \dfrac{q}{H_v}$
- Substitute the data into the equation:

$$m = \frac{4,000. \text{ J}}{\left(2259 \ \dfrac{\text{J}}{\text{g}}\right)} = \mathbf{1.771 \ g}$$

---

9. Temperature: $K = {}^\circ C + 273$

Problem

What is the Kelvin temperature equivalent of 34°C?

Solution

- Substitute the data into the equation:
  $K = 34{}^\circ C + 273 = \textbf{307 K}$

10a. Radioactive Decay: fraction remaining $= \left(\dfrac{1}{2}\right)^{\frac{t}{T}}$

10b. Radioactive Decay: number of half-life periods $= \dfrac{t}{T}$

Problem

What fraction of a sample of ^{42}K will remain unchanged after 37.2 hours of decay?

Solution

- Use Reference Table N to obtain the half-life of ^{42}K (12.4 h).
- Use equation 10b to determine the number of half-life periods:

  number of half-life periods $= \dfrac{37.2\,h}{12.4\,h} = 3.00 = \textbf{3}$

Substitute the number of half-life periods into equation 10a:

fraction remaining $= \left(\dfrac{1}{2}\right)^{3} = \dfrac{1}{8}$

That is, one-eighth of the original sample of ^{42}K remains unchanged after 37.2 hours of decay.

Regents Examinations, Answers, and Self-Analysis Charts

Examination June 2016

Chemistry—Physical Setting

PART A

Answer all questions in this part.

Directions (1–30): For *each* statement or question, write in the answer space the *number* of the word or expression that, of those given, best completes the statement or answers the question. Some questions may require the use of the *2011 Edition Reference Tables for Physical Setting/Chemistry*.

1 Which statement describes the charge of an electron and the charge of a proton?

(1) An electron and a proton both have a charge of +1.
(2) An electron and a proton both have a charge of −1.
(3) An electron has a charge of +1, and a proton has a charge of −1.
(4) An electron has a charge of −1, and a proton has a charge of +1.   1 _____

2 Which subatomic particles are found in the nucleus of an atom of beryllium?

(1) electrons and protons   (3) neutrons and protons
(2) electrons and positrons   (4) neutrons and electrons   2 _____

3 The elements in Period 4 on the Periodic Table are arranged in order of increasing

(1) atomic radius
(2) atomic number
(3) number of valence electrons
(4) number of occupied shells of electrons   3 _____

4 Which phrase describes two forms of solid carbon, diamond and graphite, at STP?

   (1) the same crystal structure and the same properties
   (2) the same crystal structure and different properties
   (3) different crystal structures and the same properties
   (4) different crystal structures and different properties   4 _____

5 Which element has six valence electrons in each of its atoms in the ground state?

   (1) Se   (3) Kr
   (2) As   (4) Ga   5 _____

6 What is the chemical name for $H_2SO_3(aq)$?

   (1) sulfuric acid   (3) hydrosulfuric acid
   (2) sulfurous acid   (4) hydrosulfurous acid   6 _____

7 Which substance is most soluble in water?

   (1) $(NH_4)_3PO_4$ (3) Ag_2SO_4
   (2) $Cu(OH)_2$ (4) $CaCO_3$ 7 _____

8 Which type of bonding is present in a sample of an element that is malleable?

   (1) ionic   (3) nonpolar covalent
   (2) metallic   (4) polar covalent   8 _____

9 Which atom has the greatest attraction for the electrons in a chemical bond?

   (1) hydrogen   (3) silicon
   (2) oxygen   (4) sulfur   9 _____

10 Which type of reaction involves the transfer of electrons?

   (1) alpha decay   (3) neutralization
   (2) double replacement   (4) oxidation-reduction   10 _____

11 A 10.0-gram sample of nitrogen is at STP. Which property will increase when the sample is cooled to 72 K at standard pressure?

(1) mass (3) density
(2) volume (4) temperature 11 _____

12 Which element is a gas at STP?

(1) sulfur (3) potassium
(2) xenon (4) phosphorus 12 _____

13 A 5.0-gram sample of Fe(s) is to be placed in 100. milliliters of HCl(aq). Which changes will result in the fastest rate of reaction?

(1) increasing the surface area of Fe(s) and increasing the concentration of HCl(aq)
(2) increasing the surface area of Fe(s) and decreasing the concentration of HCl(aq)
(3) decreasing the surface area of Fe(s) and increasing the concentration of HCl(aq)
(4) decreasing the surface area of Fe(s) and decreasing the concentration of HCl(aq) 13 _____

14 Which process is commonly used to separate a mixture of ethanol and water?

(1) distillation (3) filtration
(2) ionization (4) titration 14 _____

15 A sample of hydrogen gas will behave most like an ideal gas under the conditions of

(1) low pressure and low temperature
(2) low pressure and high temperature
(3) high pressure and low temperature
(4) high pressure and high temperature 15 _____

16 The collision theory states that a reaction is most likely to occur when the reactant particles collide with the proper

(1) formula masses (3) density and volume
(2) molecular masses (4) energy and orientation 16 _____

17 At STP, which sample contains the same number of molecules as 3.0 liters of $H_2(g)$?

   (1) 1.5 L of $NH_3(g)$ (3) 3.0 L of $CH_4(g)$
   (2) 2.0 L of $CO_2(g)$ (4) 6.0 L of $N_2(g)$ 17 _____

18 The addition of a catalyst to a chemical reaction provides an alternate pathway that

   (1) increases the potential energy of reactants
   (2) decreases the potential energy of reactants
   (3) increases the activation energy
   (4) decreases the activation energy   18 _____

19 A sample of water is boiling as heat is added at a constant rate. Which statement describes the potential energy and the average kinetic energy of the water molecules in this sample?

   (1) The potential energy decreases and the average kinetic energy remains the same.
   (2) The potential energy decreases and the average kinetic energy increases.
   (3) The potential energy increases and the average kinetic energy remains the same.
   (4) The potential energy increases and the average kinetic energy increases.   19 _____

20 Entropy is a measure of the

   (1) acidity of a sample
   (2) disorder of a system
   (3) concentration of a solution
   (4) chemical activity of an element   20 _____

21 Which element has atoms that can bond with each other to form ring, chain, and network structures?

   (1) aluminum   (3) carbon
   (2) calcium   (4) argon   21 _____

22 What is the number of electrons shared in the multiple carbon-carbon bond in one molecule of 1-pentyne?

(1) 6   (3) 3

(2) 2   (4) 8   22 _____

23 Butanal, butanone, and diethyl ether have different properties because the molecules of each compound differ in their

(1) numbers of carbon atoms

(2) numbers of oxygen atoms

(3) types of functional groups

(4) types of radioactive isotopes   23 _____

24 What occurs when a magnesium atom becomes a magnesium ion?

(1) Electrons are gained and the oxidation number increases.

(2) Electrons are gained and the oxidation number decreases.

(3) Electrons are lost and the oxidation number increases.

(4) Electrons are lost and the oxidation number decreases.   24 _____

25 Energy is required to produce a chemical change during

(1) chromatography   (3) boiling

(2) electrolysis   (4) melting   25 _____

26 The reaction of an Arrhenius acid with an Arrhenius base produces water and

(1) a salt   (3) an aldehyde

(2) an ester   (4) a halocarbon   26 _____

27 One acid-base theory defines an acid as an

(1) H^- acceptor   (3) H^+ acceptor

(2) H^- donor   (4) H^+ donor   27 _____

28 Which phrase describes the decay modes and the half-lives of K-37 and K-42?

(1) the same decay mode but different half-lives

(2) the same decay mode and the same half-life

(3) different decay modes and different half-lives

(4) different decay modes but the same half-life   28 _____

29 Which particle has a mass that is approximately equal to the mass of a proton?

  (1) an alpha particle   (3) a neutron
  (2) a beta particle   (4) a positron   29 _____

30 Which change occurs during a nuclear fission reaction?

  (1) Covalent bonds are converted to ionic bonds.
  (2) Isotopes are converted to isomers.
  (3) Temperature is converted to mass.
  (4) Matter is converted to energy.   30 _____

PART B–1

Answer all questions in this part.

Directions (31–50): For *each* statement or question, write in the answer space the *number* of the word or expression that, of those given, best completes the statement or answers the question. Some questions may require the use of the *2011 Edition Reference Tables for Physical Setting/Chemistry.*

31 Which notations represent hydrogen isotopes?

  (1) $_1^1H$ and $_1^2H$ (3) $_2^1H$ and $_3^1H$

  (2) $_1^1H$ and $_2^4H$ (4) $_1^2H$ and $_2^7H$ 31 _____

32 Naturally occurring gallium is a mixture of isotopes that contains 60.11% of Ga-69 (atomic mass = 68.93 u) and 39.89% of Ga-71 (atomic mass = 70.92 u). Which numerical setup can be used to determine the atomic mass of naturally occurring gallium?

  (1) $\dfrac{(68.93\ u + 70.92\ u)}{2}$

  (2) $\dfrac{(68.93\ u)(0.6011)}{(70.92\ u)(0.3989)}$

  (3) $(68.93\ u)(0.6011) + (70.92\ u)(0.3989)$

  (4) $(68.93\ u)(39.89) + (70.92\ u)(60.11)$ 32 _____

33 Which list of symbols represents nonmetals, only?

  (1) B, Al, Ga   (3) C, Si, Ge

  (2) Li, Be, B   (4) P, S, Cl   33 _____

34 In the formula XSO_4, the symbol X could represent the element

  (1) Al   (3) Mg

  (2) Ar   (4) Na   34 _____

35 What is the chemical formula for lead(IV) oxide?

  (1) PbO_2 (3) Pb_2O

  (2) PbO_4 (4) Pb_4O 35 _____

36 Which statement describes the general trends in electronegativity and atomic radius as the elements in Period 2 are considered in order from left to right?

  (1) Both electronegativity and atomic radius increase.
  (2) Both electronegativity and atomic radius decrease.
  (3) Electronegativity increases and atomic radius decreases.
  (4) Electronegativity decreases and atomic radius increases.   36 _____

37 What is the percent composition by mass of nitrogen in $(NH_4)_2CO_3$ (gram-formula mass = 96.0 g/mol)?

  (1) 14.6%   (3) 58.4%
  (2) 29.2%   (4) 87.5%   37 _____

38 Given the balanced equation:

$$2KI + F_2 \rightarrow 2KF + I_2$$

Which type of chemical reaction does this equation represent?

  (1) synthesis   (3) single replacement
  (2) decomposition   (4) double replacement   38 _____

39 Which formula represents a nonpolar molecule containing polar covalent bonds?

                                 N   O

H–H   O=C=O   H H H   H   H

  (1)   (2)   (3)   (4)   39 _____

40 A reaction reaches equilibrium at 100.°C. The equation and graph representing this reaction are shown below.

$$N_2O_4(g) \rightleftharpoons 2NO_2(g)$$

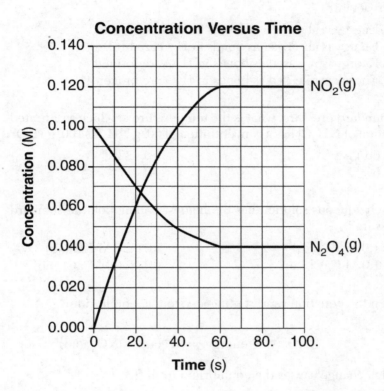

Concentration Versus Time

The graph shows that the reaction is at equilibrium after 60. seconds because the concentrations of both $NO_2(g)$ and $N_2O_4(g)$ are

(1) increasing   (3) constant

(2) decreasing   (4) zero   40 _____

41 Given the balanced equation representing a reaction:

$$2H_2O + energy \rightarrow 2H_2 + O_2$$

Which statement describes the changes in energy and bonding for the reactant?

(1) Energy is absorbed as bonds in H_2O are formed.
(2) Energy is absorbed as bonds in H_2O are broken.
(3) Energy is released as bonds in H_2O are formed.
(4) Energy is released as bonds in H_2O are broken.   41 _____

42 At standard pressure, what is the temperature at which a saturated solution of NH_4Cl has a concentration of 60. g NH_4Cl/100. g H_2O?

(1) 66°C   (3) 22°C
(2) 57°C   (4) 17°C   42 _____

43 Which aqueous solution has the highest boiling point at standard pressure?

(1) 1.0 M KCl(aq)   (3) 2.0 M KCl(aq)
(2) 1.0 M $CaCl_2$(aq)   (4) 2.0 M $CaCl_2$(aq)   43 _____

44 Given the equation representing a system at equilibrium:

$$KNO_3(s) + energy \overset{H_2O}{\rightleftharpoons} K^+(aq) + NO_3^-(aq)$$

Which change causes the equilibrium to shift?

(1) increasing pressure   (3) adding a noble gas
(2) increasing temperature   (4) adding a catalyst   44 _____

45 Which hydrocarbon is saturated?

(1) C_2H_2 (3) C_4H_6
(2) C_3H_4 (4) C_4H_{10} 45 _____

46 Which volume of 0.600 M H_2SO_4(aq) exactly neutralizes 100. milliliters of 0.300 M $Ba(OH)_2$(aq)?

(1) 25.0 mL   (3) 100. mL
(2) 50.0 mL   (4) 200. mL   46 _____

47 Given the formula for an organic compound:

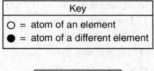

What is the name given to the group in the box?

(1) butyl   (3) methyl

(2) ethyl   (4) propyl   47 _____

48 Given the particle diagram:

Key
○ = atom of an element
● = atom of a different element

Which type of matter is represented by the particle diagram?

(1) an element

(2) a compound

(3) a homogeneous mixture

(4) a heterogeneous mixture   48 _____

49 Which substance is an electrolyte?

(1) O_2 (3) C_3H_8

(2) Xe   (4) KNO_3 49 _____

50 Which type of organic reaction produces both water and carbon dioxide?

(1) addition   (3) esterification

(2) combustion   (4) fermentation   50 _____

PART B–2

Answer all questions in this part.

Directions (51–65): Record your answers on the answer sheet provided. Some questions may require the use of the *2011 Edition Reference Tables for Physical Setting/Chemistry*.

51 Draw a Lewis electron-dot diagram for a chloride ion, Cl^-. [1]

Base your answers to questions 52 and 53 on the information below and on your knowledge of chemistry.

At STP, Cl_2 is a gas and I_2 is a solid. When hydrogen reacts with chlorine, the compound hydrogen chloride is formed. When hydrogen reacts with iodine, the compound hydrogen iodide is formed.

52 Balance the equation *on the answer sheet* for the reaction between hydrogen and chlorine, using the smallest whole-number coefficients. [1]

53 Explain, in terms of intermolecular forces, why iodine is a solid at STP but chlorine is a gas at STP. [1]

Base your answers to questions 54 and 55 on the information below and on your knowledge of chemistry.

Some properties of the element sodium are listed below.

- is a soft, silver-colored metal
- melts at a temperature of 371 K
- oxidizes easily in the presence of air
- forms compounds with nonmetallic elements in nature
- forms sodium chloride in the presence of chlorine gas

54 Identify *one* chemical property of sodium from this list.   [1]

55 Convert the melting point of sodium to degrees Celsius.   [1]

Base your answers to questions 56 through 58 on the information below and on your knowledge of chemistry.

At standard pressure, water has unusual properties that are due to both its molecular structure and intermolecular forces. For example, although most liquids contract when they freeze, water expands, making ice less dense than liquid water. Water has a much higher boiling point than most other molecular compounds having a similar gram-formula mass.

56 Explain why $H_2O(s)$ floats on $H_2O(\ell)$ when both are at 0°C.   [1]

57 State the type of intermolecular force responsible for the unusual boiling point of $H_2O(\ell)$ at standard pressure.   [1]

58 Determine the total amount of heat, in joules, required to completely vaporize a 50.0-gram sample of $H_2O(\ell)$ at its boiling point at standard pressure.   [1]

Base your answers to questions 59 and 60 on the information below and on your knowledge of chemistry.

At 1023 K and 1 atm, a 3.00-gram sample of $SnO_2(s)$ (gram-formula mass = 151 g/mol) reacts with hydrogen gas to produce tin and water, as shown in the balanced equation below.

$$SnO_2(s) + 2H_2(g) \rightarrow Sn(\ell) + 2H_2O(g)$$

59 Show a numerical setup for calculating the number of moles of $SnO_2(s)$ in the 3.00-gram sample.   [1]

60 Determine the number of moles of $Sn(\ell)$ produced when 4.0 moles of $H_2(g)$ is completely consumed.   [1]

Base your answers to questions 61 and 62 on the information below and on your knowledge of chemistry.

The incomplete data table below shows the pH value of solutions A and B and the hydrogen ion concentration of solution A.

Hydrogen Ion and pH Data for HCl(aq) Solutions

HCl(aq) Solution	Hydrogen Ion Concentration (M)	pH
A	1.0×10^{-2}	2.0
B	?	5.0

61 State the color of methyl orange in a sample of solution A.   [1]

62 Determine the hydrogen ion concentration of solution B.   [1]

Base your answers to questions 63 through 65 on the information below and on your knowledge of chemistry.

A sample of helium gas is placed in a rigid cylinder that has a movable piston. The volume of the gas is varied by moving the piston, while the temperature is held constant at 273 K. The volumes and corresponding pressures for three trials are measured and recorded in the data table below. For each of these trials, the product of pressure and volume is also calculated and recorded. For a fourth trial, only the volume is recorded.

**Pressure and Volume Data for
a Sample of Helium Gas at 273 K**

Trial Number	Pressure (atm)	Volume (L)	P × V (L•atm)
1	1.000	0.412	0.412
2	0.750	0.549	0.412
3	0.600	0.687	0.412
4	?	1.373	?

63 State evidence found in the data table that allows the product of pressure and volume for the fourth trial to be predicted.   [1]

64 Determine the pressure of the helium gas in trial 4.   [1]

65 Compare the average distances between the helium atoms in trial 1 to the average distances between the helium atoms in trial 3.   [1]

PART C

Answer all questions in this part.

Directions (66–85): Record your answers on the answer sheet provided. Some questions may require the use of the *2011 Edition Reference Tables for Physical Setting/Chemistry*.

Base your answers to questions 66 through 69 on the information below and on your knowledge of chemistry.

Potassium phosphate, K_3PO_4, is a source of dietary potassium found in a popular cereal. According to the Nutrition-Facts label shown on the boxes of this brand of cereal, the accepted value for a one-cup serving of this cereal is 170. milligrams of potassium. The minimum daily requirement of potassium is 3500 milligrams for an adult human.

66 Identify *two* types of chemical bonding in the source of dietary potassium in this cereal.   [1]

67 Identify the noble gas whose atoms have the same electron configuration as a potassium ion.   [1]

68 Compare the radius of a potassium ion to the radius of a potassium atom.   [1]

69 The mass of potassium in a one-cup serving of this cereal is determined to be 172 mg. Show a numerical setup for calculating the percent error for the mass of potassium in this serving.   [1]

Base your answers to questions 70 and 71 on the information below and on your knowledge of chemistry.

During photosynthesis, plants use carbon dioxide, water, and light energy to produce glucose, $C_6H_{12}O_6$, and oxygen. The reaction for photosynthesis is represented by the balanced equation below.

$$6CO_2 + 6H_2O + \text{light energy} \rightarrow C_6H_{12}O_6 + 6O_2$$

70 Write the empirical formula for glucose.   [1]

71 State evidence that indicates photosynthesis is an endothermic reaction.   [1]

Base your answers to questions 72 through 74 on the information below and on your knowledge of chemistry.

Fireworks that contain metallic salts such as sodium, strontium, and barium can generate bright colors. A technician investigates what colors are produced by the metallic salts by performing flame tests. During a flame test, a metallic salt is heated in the flame of a gas burner. Each metallic salt emits a characteristic colored light in the flame.

72 Explain why the electron configuration of 2-7-1-1 represents a sodium atom in an excited state.   [1]

73 Explain, in terms of electrons, how a strontium salt emits colored light.   [1]

74 State how bright-line spectra viewed through a spectroscope can be used to identify the metal ions in the salts used in the flame tests.   [1]

Base your answers to questions 75 through 77 on the information below and on your knowledge of chemistry.

The unique odors and flavors of many fruits are primarily due to small quantities of a certain class of organic compounds. The equation below represents the production of one of these compounds.

$$
\begin{array}{cccc}
\underset{\text{Reactant 1}}{\overset{\displaystyle \text{H}\ \ \text{H}}{\text{H}-\overset{\text{H}}{\underset{\text{H}}{\text{C}}}-\overset{\text{H}}{\underset{\text{H}}{\text{C}}}-\text{OH}}} +
\underset{\text{Reactant 2}}{\overset{\displaystyle \text{O}}{\text{H}-\overset{\|}{\text{C}}-\text{O}-\text{H}}} \longrightarrow
\underset{\text{Product 1}}{\overset{\displaystyle \text{O}\ \ \ \ \text{H}\ \ \text{H}}{\text{H}-\overset{\|}{\text{C}}-\text{O}-\overset{\text{H}}{\underset{\text{H}}{\text{C}}}-\overset{\text{H}}{\underset{\text{H}}{\text{C}}}-\text{H}}} +
\underset{\text{Product 2}}{\text{HOH}}
\end{array}
$$

75 Show a numerical setup for calculating the gramformula mass for reactant 1. [1]

76 Explain, in terms of molecular polarity, why reactant 2 is soluble in water. [1]

77 State the class of organic compounds to which product 1 belongs. [1]

Base your answers to questions 78 through 81 on the information below and on your knowledge of chemistry.

A student develops the list shown below that includes laboratory equipment and materials for constructing a voltaic cell.

Laboratory Equipment and Materials

- a strip of zinc
- a strip of copper
- a 250-mL beaker containing 150 mL of 0.1 M zinc nitrate
- a 250-mL beaker containing 150 mL of 0.1 M copper(II) nitrate
- wires
- a voltmeter
- a switch
- a salt bridge

78 State the purpose of the salt bridge in the voltaic cell. [1]

79 Complete and balance the half-reaction equation *on your answer sheet* for the oxidation of the Zn(s) that occurs in the voltaic cell. [1]

80 Compare the activities of the two metals used by the student for constructing the voltaic cell. [1]

81 Identify *one* item of laboratory equipment required to build an electrolytic cell that is *not* included in the list. [1]

Base your answers to questions 82 through 85 on the information below and on your knowledge of chemistry.

In 1896, Antoine H. Becquerel discovered that a uranium compound could expose a photographic plate wrapped in heavy paper in the absence of light. It was shown that the uranium compound was spontaneously releasing particles and high-energy radiation. Further tests showed the emissions from the uranium that exposed the photographic plate were *not* deflected by charged plates.

82 Identify the highly penetrating radioactive emission that exposed the photographic plates. [1]

83 Complete the nuclear equation *on your answer sheet* for the alpha decay of U-238. [1]

84 Determine the number of neutrons in an atom of U-233. [1]

85 Identify the type of nuclear reaction that occurs when an alpha or a beta particle is spontaneously emitted by a radioactive isotope. [1]

Answer Sheet June 2016

Chemistry—Physical Setting

PART B–2

51

52 _____ $H_2(g)$ + _____ $Cl_2(g) \rightarrow$ _____ $HCl(g)$

53 _____

_____

_____

54 _____

_____

55 _____ °C

56 _____

_____

_____

57 _____

58 _____ **J**

59

60 _____ **mol**

61 _____

62 _____ **M**

63 _____

_____

_____

64 _____ **atm**

65 _____

_____

_____

PART C

66 _____ and _____

67 _____

68 _____

_____

69

70 _____

71 _____

_____

_____

72 _____

73 _____

74 _____

75

76 _____

77 _____

78 _____

_____

79 $Zn(s) \rightarrow$ _____ + _____

80 _____

_____

_____

81 _____

82 _____

83 $^{238}_{92}U \rightarrow \, ^{4}_{2}He +$ _____

84 _____

85 _____

Answers
June 2016

Chemistry—Physical Setting

Answer Key

PART A

1. 4	7. 1	13. 1	19. 3	25. 2
2. 3	8. 2	14. 1	20. 2	26. 1
3. 2	9. 2	15. 2	21. 3	27. 4
4. 4	10. 4	16. 4	22. 1	28. 3
5. 1	11. 3	17. 3	23. 3	29. 3
6. 2	12. 2	18. 4	24. 3	30. 4

PART B–1

31. 1	35. 1	39. 2	43. 4	47. 2
32. 3	36. 3	40. 3	44. 2	48. 2
33. 4	37. 2	41. 2	45. 4	49. 4
34. 3	38. 3	42. 1	46. 2	50. 2

PART B–2 and **PART C**. *See* **Answers Explained**.

Answers Explained

PART A

1. **4** The charges of all subatomic particles should be committed to memory, but *just to be sure*, you can consult Reference Table *O*, which shows the symbols, mass numbers, and charges for electrons, neutrons, and protons.

2. **3** The nucleus contains only protons and neutrons. Electrons are found in the electron cloud surrounding the nucleus.

3. **2** The Periodic Table found in the Reference Tables illustrates that the elements are organized in order of increasing atomic number.

WRONG CHOICES EXPLAINED:
(1) Atomic radius data is provided in Reference Table *S*. The atomic radius decreases in Period 4 of the Periodic Table. This happens because as elements have more protons in their nuclei, the stronger positive charge shrinks the size of the electron cloud.
(3) The number of valence electrons is the last number in the electron configurations provided in the Periodic Table of the Elements. The number of valence electrons does not increase regularly for the transition metals in Period 4.
(4) The number of occupied shells of electrons is 4 for all Period 4 elements. Notice that there are four values present in the electron configurations for Period 4 elements.

4. **4** Diamond and graphite are two *allotropes* of carbon, meaning that they are naturally occurring forms of carbon that have different structural arrangements. Diamond is arranged in a three-dimensional tetrahedral structure, and graphite is arranged in flat sheets, as shown in the illustrations below. Having different structures causes these substances to have different chemical and physical properties.

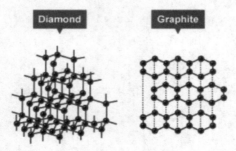

Diamond   Graphite

5. **1** Valence electrons are the electrons in the outer shell, or energy level, of an atom. These are the electrons involved in making chemical bonds with other atoms. The last number in an element's electron configuration is the number of valence electrons for that element. Looking at the Periodic Table of the Elements in the Reference Tables, the electron configuration for selenium (Se) is 2-8-18-6.

WRONG CHOICES EXPLAINED:
(2) The electron configuration for arsenic (As) is 2-8-18-5, so As has 5 valence electrons.
(3) The electron configuration for krypton (Kr) is 2-8-18-8, so Kr has 8 valence electrons.
(4) The electron configuration for gallium (Ga) is 2-8-18-3, so Ga has 3 valence electrons.

6. **2** If a compound's chemical formula begins with hydrogen, there is a good bet that the compound is an acid. One exception is water, H_2O, but even water does at times act as an acid. The chemical name for H_2SO_3 (aq) can be found in Reference Table K.

7. **1** To learn whether or not a compound is soluble in water, consult Reference Table F. The ammonium cation, NH_4^+, is listed in the first column of the table, which means that all compounds containing the ammonium ion are soluble in water, with no exceptions.

WRONG CHOICES EXPLAINED:
(2) The hydroxide ion (OH^-) forms insoluble compounds in water, and Cu^{2+} is not an exception to this insolubility.
(3) The sulfate ion (SO_4^{2-}) forms soluble compounds on water *most of the time*, but Ag^+ is listed as an exception to this solubility (making Ag_2SO_4 insoluble in water).
(4) The carbonate ion (CO_3^{2-}) forms insoluble compounds in water, and Ca^{2-} is not an exception to this insolubility.

8. **2** Malleability, the ability of a material to be hammered or pressed into thin sheets, is a physical property of metallic substances. Metals are also ductile (able to be stretched into wires), shiny (having metallic luster), and highly conductive. These properties all arise from the mobile sea of electrons, the key feature in metallic bonding.

9. **2** This is a "definition question" that requires the use of the Reference Tables. Electronegativity measures an atom's attraction for the electrons in a chemical bond. Electronegativity values can be found in Reference Table S. Of all the choices, oxygen's electronegativity of 3.4 is the highest.

10. **4** This is a "definition question!" An oxidation-reduction reaction (nick-named a "redox" reaction) is defined as a chemical reaction that involves the transfer of electrons.

WRONG CHOICES EXPLAINED:
(1) Alpha decay is a form of radioactive decay, which is a type of nuclear reaction.
(2) A double replacement reaction is a reaction between two ionic compounds in which the cations and anions switch places, forming two new ionic compounds.
(3) A neutralization reaction takes place when an acid reacts with a base to form water and a salt.

11. **3** Reference Table S shows that nitrogen boils at a temperature of 77 K. If a sample of nitrogen gas at STP is cooled to 72 K, it will undergo a phase change from gas to liquid. Because particles are closer together in the liquid phase, the liquid phase is always more dense than the gas phase for any substance.

WRONG CHOICES EXPLAINED:
(1) Mass is unaffected by temperature change.
(2) The volume of a gas is directly related to temperature, so cooling a sample of gas will cause the volume to decrease. In addition, the phase change to the liquid state will cause an even more dramatic reduction in volume.
(4) Cooling from 273 K (standard temperature) to 72 K constitutes a decrease in temperature.

12. **2** Xenon is a noble gas, which by definition is a gas at room temperature. Data in Reference Table S shows that xenon boils at a temperature of 165 K, which is well below room temperature. All of the other answer choices are solids at room temperature.

13. **1** The rate of a chemical reaction increases when the number of collisions between reactant particles increases. If the concentration of HCl(aq) increases, there will be more HCl particles available to collide with Fe(s). Likewise, if the surface area of the Fe(s) increases by using smaller pieces of Fe(s), there will be more exposed surface available for collisions with HCl.

14. **1** Ethanol is a liquid organic compound that is completely soluble in water. Distillation is used to separate the components of a liquid mixture based on their different boiling points.

15. **2** Real gases behave more like ideal gases when temperature is high and pressure is low. These conditions minimize any possible attractions between gas particles, allowing the gas to adhere more closely to the principles of the kinetic molecular theory of gases.

16. **4** Particles involved in a reaction may collide frequently, but for a reaction to occur the collision must be an "effective" collision. Effective collisions have both sufficient energy and proper orientation.

17. **3** Equal volumes of gas at the same conditions of temperature and pressure contain the same number of particles. This relationship is known as Avogadro's Law. For example, the molar volume of a gas at STP = 22.4 L/mol no matter what type of gas you have.

18. **4** A catalyst lowers the activation energy of a chemical reaction as shown in the energy diagram below.

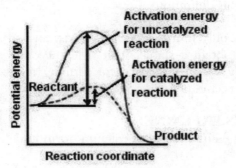

Since less energy is needed for a reaction to occur, more collisions will be effective and the reaction rate will increase.

19. **3** Temperature is directly related to kinetic energy, which is the energy of motion of particles in any sample of matter. During phase changes, such as the boiling of water, the temperature does not change, as shown in the diagram below.

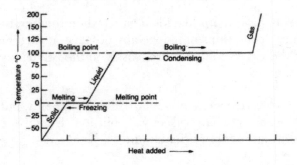

This means that the kinetic energy is constant during boiling. Heat energy added during boiling causes water molecules to overcome their attractions to each other and spread apart, forming water vapor. Increasing the distance between water molecules increases their potential energy.

20. **2** Entropy is defined as the degree of chaos, randomness, or disorder of a system. Remember this definition!

21. **3** Carbon has only four valence electrons, and so atoms of carbon will form four bonds to obtain a stable octet of eight valence electrons. Because carbon forms four bonds, it is capable of forming many different structures, including rings, chains, and networks.

WRONG CHOICES EXPLAINED:

(1), (2) Metals, like aluminum and calcium, form ionic compounds that have lattice structures.

(4) Argon is a very stable, monatomic noble gas that rarely, if ever, forms compounds.

22. **1** Reference Table Q shows that pentyne, a member of the alkyne family of hydrocarbons, contains a triple carbon-carbon bond. A single covalent bond consists of a *shared pair* of electrons. A triple bond would contain *three shared pairs* of electrons, for a total of six electrons shared between these two carbon atoms.

23. **3** Butanal is an aldehyde, butanone is a ketone, and diethyl ether is an ether. These compounds contain different organic functional groups, as shown on Reference Table R.

24. **3** The Periodic Table of the Elements shows that the magnesium ion has a charge of +2. Magnesium must lose two electrons to become the magnesium ion, Mg^{+2}. The oxidation number of simple ions is equal to the charge on the ion, so the magnesium ion has an oxidation number of +2. Since a magnesium atom has an oxidation number of 0, the oxidation number increases when the ion forms.

25. **2** Electrolysis, performed in an electrolytic cell, requires an input of energy from a power source to force electrons to move in a nonspontaneous direction.

WRONG CHOICES EXPLAINED:
(1) Chromatography is a method for separating liquid or gas mixtures. The separation of mixtures is a physical change, not a chemical change.
(3), (4) Boiling and melting are phase changes. Phase changes are reversible physical changes since no new substance forms.

26. **1** An example of an acid-base neutralization reaction is shown below:

$$HCl \quad + \quad NaOH \quad \rightarrow \quad H_2O \quad + \quad NaCl$$
$$\text{(acid)} \qquad \text{(base)} \qquad \text{(water)} \qquad \text{(salt)}$$

The products are always water and a salt in neutralization reactions between Arrhenius acids and bases. It might help to think of these reactions as a special type of double replacement reaction in which the ions trade places.

27. **4** The Brønsted-Lowry theory of acids and bases defines acids as proton (H^+) donors and bases as proton (H^+) acceptors. Remember this definition!

28. **3** Consult Reference Table N for information about decay modes and half-lives. K-37 and K-42 differ both in decay modes and half-lives.

29. **3** Consult Reference Table O for information about nuclear particles. Protons and neutrons both have a mass of 1 atomic mass unit (amu).

30. **4** Nuclear fission releases a tremendous amount of energy. This energy results from the conversion of a small amount of matter into energy according to Einstein's famous equation $E = mc^2$.

PART B–1

31. **1** Isotopes are different versions of an element with different numbers of neutrons. All hydrogen atoms must have an atomic number (the bottom number) of 1. Different isotopes of hydrogen will have different mass numbers (the top number) reflecting the different number of neutrons they contain.

WRONG CHOICES EXPLAINED:
  (2), (4) Hydrogen atoms can never have an atomic number of 2.
  (3) The atomic number, which is 1 for all atoms of hydrogen, must be written as the *bottom number* in standard isotopic notation.

32. **3** The average atomic mass of any element is calculated by taking the weighted average of all of its naturally occurring isotopes. The percent abundance of each naturally occurring isotope is multiplied by its atomic mass, resulting in an average that reflects how abundant each isotope is in nature.

33. **4** Nonmetals, with the exception of hydrogen, are located to the right of the staircase dividing line on the Periodic Table.

WRONG CHOICES EXPLAINED:
  (1) Aluminum (Al) and gallium (Ga) are metals (on the left side of the dividing line), and boron (B) is a metalloid. [The six metalloids must be memorized. They are B, Si, As, Te, Sb, and Ge. Look at the positions of these elements on the Periodic Table and visually memorize the physical pattern that they make.]
  (2) Lithium (Li) and beryllium (Be) are metals, and boron (B) is a metalloid.
  (3) Carbon (C) is a nonmetal, but silicon (Si) and germanium (Ge) are metalloids.

34. **3** You can find the charge of the sulfate ion, SO_4^{2-}, on Reference Table *E*. Compounds have no overall charge, so the -2 charge of the sulfate ion must be balanced out by a single ion having a charge of $+2$. Charges for ions of single elements can be found on the Periodic Table provided in your Reference Tables.

35. **1** A Roman numeral is provided in the names of many ionic compounds to indicate the charges on the metal ions in these compounds. The formula for lead(IV) oxide contains the Roman numeral IV, which is equal to 4. Looking at the Periodic Table of the Elements, we confirm that $+4$ is a possible charge for lead. Since all compounds are neutral in charge, and remembering that the oxide ion has a charge of -2 (also found on the Periodic Table), we can put together the following information:

$$Pb^{+4} \quad O^{-2}$$

To finish the chemical formula, two oxide ions are needed to offset the +4 charge on lead. The correct formula, PbO_2 is the only answer choice that is balanced in charge.

36. **3** Since electronegativity and atomic radius data for most of the elements is listed on Reference Table S, there is never any reason to guess on periodic trends for these properties. Looking at the elements in Period 2 of the Periodic Table, electronegativity *increases* and atomic radius *decreases* as the elements are considered in order from left to right. These trends are caused by increasing nuclear charge, which draws the electron cloud in closer (shrinks that atomic radius) and causes the nucleus of each successive element to be more attractive to electrons in chemical bonds (electronegativity).

37. **2** Using the percent composition formula on Reference Table T:

$$\% \text{ composition by mass} = \frac{\text{mass of part}}{\text{mass of whole}} \times 10$$

$$\% \text{ composition by mass of nitrogen in } (NH_4)_2CO_3$$

$$= \frac{(14.0 \text{ g/mol})(2)}{96.0 \text{ g/mol}} \times 100 = \frac{28.0 \text{ g/mol}}{96.0 \text{ g/mol}} \times 100 \approx 29.2\%$$

38. **3** A single replacement reaction takes place when a more reactive element replaces a less reactive element in a compound. Single replacement reactions can be recognized by the presence of *uncombined elements on each side of the equation*.

WRONG CHOICES EXPLAINED:

(1) Synthesis reactions follow the pattern $A + B \rightarrow AB$. Two reactants combine to form one product.

(2) Decomposition reactions follow the pattern $AB \rightarrow A + B$. One reactant breaks down to form two or more products.

(4) Double replacement reactions follow the pattern $AB + CD \rightarrow AD + CB$. Two compounds appear on both sides of the reaction, and the reactant cations switch places, forming the products.

39. **2** Bonds between two different nonmetal elements with different electronegativities are polar covalent bonds. Although the double bonds between C and O are polar covalent, the symmetry of the CO_2 molecule cancels the polarity of the bonds. The result is an overall nonpolar molecule.

WRONG CHOICES EXPLAINED:

(1) H−H contains a nonpolar covalent bond between two identical nonmetal atoms.

(3), (4) These molecules are not symmetrical. The polarity in their polar covalent bonds does not cancel out, so these will be polar molecules.

40. **3** When chemical equilibrium is established in a reaction system, the rates of the forward and reverse reactions are equal. Since reactant and product molecules are forming at equal rates, the concentrations of all molecules remain constant at equilibrium.

41. **2** Energy is required to break apart chemical bonds, in just the same way as it takes energy to split apart two magnets that are sticking together. The reaction shows water molecules decomposing into their constituent elements. Energy must be absorbed into the water molecules to break apart the bonds.

42. **1** Use Reference Table G to find this answer. Note that the information given, 60. g NH_4Cl in 100. g H_2O, is the y-axis coordinate. Look across to find the temperature that corresponds to this value on the solubility curve for NH_4Cl.

43. **4** Boiling point elevation is one of the colligative properties of solutions. The presence of solute particles causes water to boil at a higher temperature than it normally would, and the more concentrated the solute particles, the higher the boiling temperature becomes. A 2.0 M solution of $CaCl_2$ would yield the highest concentration of ions in solution.

44. **2** Le Châtelier's principle states that when a stress is applied to a system at equilibrium, the system will shift (react) to counteract the stress and restore equilibrium. Increasing the temperature on the reaction system shown would cause a shift to the right to consume the excess heat and bring the temperature back down again, thus restoring equilibrium.

WRONG CHOICES EXPLAINED:

(1), (3) Both refer to increasing the pressure within the reaction system. Pressure is only a stress on equilibrium if one of the reactants or products is a gas. There are no gases in the reactions shown.

(4) Catalysts increase the rate of *both the forward and reverse reactions* in a system at equilibrium, but the position of equilibrium is unaffected.

45. Saturated hydrocarbons contain only single covalent bonds between carbon atoms. Memorize this definition! Reference Table Q shows that alkanes have only single C−C bonds, so the answer will follow the general formula C_nH_{2n+2}.

46. **2** Use the titration formula provided on Reference Table T of your Reference Tables:

$$M_AV_A = M_BV_B$$

in which M_A = the molarity of H^+, V_A = volume of acid, M_B = molarity of OH^-, and V_B = volume of base.

In this case, since H_2SO_4 is a *diprotic* acid, two H^- ions will be released for every molecule that dissociates. We must double the molarity of the H_2SO_4 given to find M_A, the molarity of the H^- ions.

$$M_A = 2(0.600 \text{ M}) = 1.200 \text{ M}$$

Likewise, $Ba(OH)_2$ is a *dihydroxy* base and will release two moles of OH^- ions for every mole of base that dissociates. We must double the molarity of the $Ba(OH)_2$ given to find M_B, the molarity of OH^- ions.

$$M_B = 2(0.300 \text{ M}) = 0.600 \text{ M}$$

Substituting the values we have just determined and the other information given:

$$(1.200 \text{ M})(V_A) = 0.600 \text{ M} (100. \text{ mL})$$
$$V_A = \frac{(0.600 \text{ M})(100. \text{ mL})}{1.200 \text{ M}} = 50.0 \text{ mL}$$

47. **2** The two-carbon branch shown in the box is named following the IUPAC system for naming hydrocarbons. There are two carbons in the branch, so the prefix, found in Reference Table P is "eth-." The suffix for hydrocarbon branches that are not part of the longest chain of carbon atoms is "-yl."

48. **2** A compound consists of two or more elements that are chemically bonded together.

49. **4** The three classes of compounds that are known as electrolytes are acids, bases, and salts (ionic compounds). All of these compounds dissociate into ions in aqueous solution, and the ions formed cause the resulting solution to be capable of conducting electricity. KNO_3 is an ionic compound made from K^+ ions and NO_3^- ions.

50. **2** The general reaction for complete combustion is

$$C_xH_y + O_2 \rightarrow CO_2 + H_2O.$$

WRONG CHOICES EXPLAINED:
(1) Addition reactions typically produce halocarbons.
(3) Esterification reactions produce esters.
(4) Fermentation reactions produce alcohol and carbon dioxide.

PART B–2

[All questions in Part B–2 are worth 1 point.]

51. Lewis electron-dot diagrams for ions are drawn to show the valence electrons in the ion *and* the charge of the ion. Chlorine, as an element, has 7 valence electrons, but the chloride ion contains one additional electron, giving it a charge of −1, as shown in the correct diagram below. The convention is to always draw brackets around the octet of valence electrons for nonmetal ions, showing that any extra electrons have been *transferred completely* to the ion and will not be shared with another atom.

52. The balanced equation (using smallest whole-number coefficients) is

$$\underline{}H_2(g) + \underline{}Cl_2(g) \rightarrow \underline{2}\,HCl(g)$$

The basic idea when balancing chemical equations can be thought of as "Atoms In = Atoms Out." Since there are two atoms of hydrogen and chlorine on the left of the arrow in the equation above, there must be two atoms of each of these elements on the right side also.

(Note: Credit was granted even if coefficients of "1" were written on the reactant side of the equation.)

53. For iodine to be a solid at STP, its melting temperature must be greater than 0 °C, while chlorine must have a melting point below 0 °C. Iodine has a higher melting point because **it has stronger intermolecular forces between its molecules than chlorine does**.

54. The chemical properties of any substance are observed when the substance undergoes a chemical reaction. From the list of properties given, there are three correct answers: **sodium oxidizes easily in the presence of air, sodium forms compounds with nonmetallic elements in nature, and sodium forms sodium chloride in the presence of chlorine gas**. These properties all involve chemical reactions; the other properties are physical properties of the element sodium and do not involve chemical reactions.

55. The last equation on Reference Table T is the equation used to convert from the Celsius to the Kelvin scale and vice versa.

$$K = °C + 273$$

According to the information given in this problem, sodium melts at 371 K. The formula above can be rearranged to solve for degrees Celsius:

$$°C = K - 273$$

Plugging in:

$$°C = 371 - 273 = \textbf{98°C}$$

56. The answer to this question is stated directly in the information about water provided, making this a "careful reading" question. The second sentence states that **"when [water freezes, it] expands, making ice [$H_2O(s)$] less dense than liquid water [$H_2O(\ell)$]."**

57. **Hydrogen bonding** is an extremely strong type of intermolecular force. Water molecules are strongly attracted to each other because of hydrogen bonding, causing water to have a much higher boiling point than would be expected for such a small molecule. Hydrogen bonding occurs in compounds that have one or more hydrogen atoms *directly bonded to* nitrogen, oxygen, or fluorine atoms, which are all highly electronegative.

58. The last page of the Reference Tables, Reference Table *T*, lists the formula to find the heat required to vaporize any pure substance in the liquid phase:

$$q = mH_v$$

where q = heat, m = mass, and H_v is the heat of vaporization.

The mass is given as 50.0 grams, and the heat of vaporization for liquid water, found on Reference Table *B*, is 2260 J/g. Plug these values into the formula and solve:

$$q = (50.0 \text{ g})(2260 \text{ J/g}) = \mathbf{113{,}000 \text{ J}}$$

59. Use the Mole Calculations formula found on Reference Table *T*:

$$\text{number of moles} = \frac{\text{given mass}}{\text{gram-formula mass}}$$

According to the information given in this problem, the sample mass is 3.00 grams, and the gram-formula mass of SnO_2 (s) is 151 g/mol.

To finish the problem, substitute the known values into the equation:

$$\textbf{number of moles } SnO_2 = \frac{\textbf{3.00 g}}{\textbf{151 g/mol}}$$

(NOTE: This problem only asks for a numerical setup. You are NOT required to solve for the answer. If you DO solve for the actual number of moles and you solve incorrectly, you will LOSE the point for this question even if you have shown the correct numerical setup. If you wish to solve for the number of moles to check that your setup is correct, do so on scrap paper.)

60. All coefficients in chemical reactions represent particles or moles. Mole ratios can be used to find the number of moles of any substance in a chemical equation if the moles of another substance are known. Two methods of setting up this problem are:

$$\frac{4.0 \ \cancel{\text{mol } H_2(g)}}{1} \times \frac{1 \ \text{mol Sn}(\ell)}{2 \ \cancel{\text{mol } H_2(g)}} = \textbf{2.0 mol Sn}(\boldsymbol{\ell})$$

OR

$$\frac{2 \ \text{mol } H_2(g)}{1 \ \text{mol Sn}(\ell)} = \frac{4.0 \ \text{mol } H_2(g)}{x \ \text{mol Sn}(\ell)}$$

$$\boldsymbol{x = \textbf{2.0 mol Sn}(\boldsymbol{\ell})}$$

61. The information presented in Reference Table M can be interpreted as follows: below a pH of 3.1, solutions would be red in the presence of methyl orange indicator, and above a pH of 4.4, solutions would appear yellow. Solutions with a pH between 3.1 and 4.4 would appear orange (a blend of the two colors).

This means that, according to Reference Table M, methyl orange indicator would appear **red** in a solution that has a pH of 2.0.

62. pH is defined as the *negative of the base 10 exponent* of the hydrogen (hydronium) ion concentration in any aqueous solution. For example, a solution that has a pH of 4 has an H^+ concentration of 1.0×10^{-4} M. Since the pH of solution B is 5.0, the molarity of solution B is $\mathbf{1.0 \times 10^{-5}}$ **M**.

63. Since $\mathbf{P \times V}$ **is constant at 0.412 L · atm for the first three trials**, a prediction can be made that the value will remain constant for the fourth trial.

64. Using the assumption that P × V for Trial 4 will be 0.412 L · atm, the pressure of the helium gas in this trial can be calculated.
We know that:

$$P \times V = 0.412 \ L \cdot atm$$

Plugging in the volume of the gas in Trial 4:

$$P \times 1.373 \ L = 0.412 \ L \cdot atm$$

To solve, divide both sides by 1.373 L:

$$\frac{(P \times \cancel{1.373L})}{\cancel{1.373L}} = \frac{0.412 \, \cancel{L} \cdot atm}{1.373 \, \cancel{L}} = \textbf{0.300 atm (or 0.3 atm or 0.30 atm)}$$

65. When faced with problem-solving in chemistry, it often helps to draw a diagram of the system, or, at the very least, have a good picture in your mind. In this problem, a sample of gas is placed in a rigid cylinder with a movable piston, as shown in the diagram below.

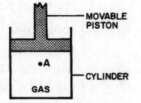

If pressure is put on the piston, the gas sample will be squeezed into a smaller volume. Since the pressure is greater in trial 1, **the average distance between the helium atoms will be smaller in trial 1 than in trial 3**.

PART C

[All questions in Part C are worth 1 point.]

66. Potassium phosphate, K_3PO_4, is an ionic compound that consists of the potassium ion and the polyatomic phosphate ion, which contains the nonmetal elements phosphorus and oxygen. Remember, ionic bonds form when metals bond to nonmetals. However, within the phosphate ion, PO_4^{-3}, the phosphorus is *covalently* bonded to four different oxygen atoms. Covalent bonds form when nonmetal atoms bond to each other. Therefore, there are both **ionic and covalent** bonds present in the compound K_3PO_4.

67. Using the Periodic Table of the Elements provided in the Reference Tables, we learn that potassium makes ions having a charge of +1. A potassium atom must *lose* one electron to become an ion with a +1 charge. Neutral potassium atoms have 19 protons and 19 electrons, but after losing an electron, potassium ions have only 18 electrons, which is the number of electrons found in neutral atoms of the element **argon (Ar)**.

68. A simplified diagram of a potassium atom is shown below:

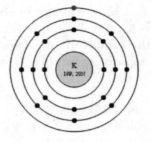

When a potassium atom becomes a potassium ion, K^{+1}, the valence electron in the 4th principal energy level is lost. Since the ion has no electrons occupying the 4th principal energy level, **the radius of the potassium ion is smaller than the radius of the potassium atom**.

69. As in question 59, you are required to write *only* a numerical setup to answer this question. See the note for question 59 concerning numerical setup problems.

From the information provided, the accepted value of potassium in a one-cup serving of cereal is 170. mg. The formula for percent error is found in Reference Table *T*:

$$\% \text{ error} = \frac{\text{measured value} - \text{accepted value}}{\text{accepted value}} \times 100$$

Plug in the measured value of 172 mg given in this problem and the accepted value of 170. mg:

$$\% \text{ error} = \frac{172 \text{ mg} - 170. \text{ mg}}{170. \text{ mg}} \times 100$$

70. Empirical formulas give the smallest whole-number ratio of atoms in a compound, allowing one to see at a glance the proportion of atoms in the compound. An empirical formula reduces the subscripts in a compound formula into "lowest terms." Since the chemical formula of glucose is $C_6H_{12}O_6$, its empirical formula is **CH_2O**.

71. An endothermic reaction absorbs energy from the surroundings. The reading passage for this question states that "plants use carbon dioxide, water, and light energy…" This means that **light energy is absorbed by plants during photosynthesis**. Another way to answer this question is to note that **the energy term is on the left side of the equation** for photosynthesis. If energy is on the left side of the equation, it is consumed, just like the reactants, during the course of the reaction.

72. The Periodic Table in the Reference Tables shows that the ground-state electron configuration of a sodium atom is 2-8-1. A sodium atom would be in an excited state if its configuration were 2-7-1-1 because this configuration shows that **a second shell electron has moved to the fourth shell**. Another acceptable answer is that in the electron configuration 2-7-1-1, **not all of the 11 electrons are in the lowest possible energy levels**.

73. Since the question specifically states that students must answer *in terms of electrons,* you must specifically mention electrons in your response. Electrons within strontium atoms can move to higher energy levels within the atom when heat energy is absorbed from a gas burner during a flame test. **This energy is emitted as photons of light with specific wavelengths and colors as electrons drop back to their lower energy, ground-state positions**.

74. The color of light produced in a flame test can be separated into a bright line spectrum using a spectroscope. Each element produces a unique bright line spectrum. **By comparing the bright line spectrum observed through the spectroscope to spectra for known elements**, the identity of the metal ions in the salt can be established.

75. As in questions 59 and 69, you are required to write *only* a numerical setup to answer this question. See the note for question 59 concerning numerical setup problems.

The gram-formula mass, also known as the molar mass, can be calculated by adding the molar masses of all of the elements in a compound. First, notice that there are two carbon atoms, six hydrogen atoms, and one oxygen atom in Reactant 1. You can find the molar masses of each element in the Periodic Table provided in your Reference Tables.

Algebraically, the gram-formula mass can be expressed as:

$$\left[2 \times (\text{molar mass of carbon})\right] + \left[6 \times (\text{molar mass of hydrogen})\right]$$
$$+ \left[1 \times (\text{molar mass of oxygen})\right]$$

You are asked, however, for a *numerical* setup, so substituting for the molar masses of C, H, and O, the expression becomes:

$$(2 \times 12.0) + (6 \times 1.0) + (1 \times 16.0)$$

76. Water molecules are polar, meaning that water molecules have an end with a partial positive charge and an end with a partial negative charge as shown in the diagram below.

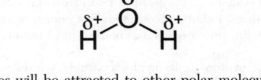

Water molecules will be attracted to other polar molecules since opposite charges will attract. **Reactant 2 will dissolve in water because molecules of both compounds are polar**.

77. Reference Table R is useful for classifying organic compounds. **Product 1 is an ester** because it matches the general formula for esters shown below and in Table R.

$$\overset{\displaystyle O}{\overset{\|}{R-C-O-R'}}$$

78. In a voltaic cell, **a salt bridge allows ions to flow into both compartments to preserve the balance of charge as the cell operates**.

79. Oxidation is defined as the loss of electrons. In this voltaic cell, the solid zinc electrode will oxidize (lose electrons), forming Zn^{+2} ions. This oxidation can be represented by the following half-reaction:

$$Zn(s) \rightarrow Zn^{2+} + 2e^-$$

80. Reference Table J lists metals from the most active metals at the top to the least active metals at the bottom. A more active metal will oxidize in the presence of a less active metal. Since zinc is higher than copper on Table J, **zinc (Zn) is more active than copper (Cu)**.

81. This question references the list of equipment needed for a *voltaic* cell, which utilizes a spontaneous redox reaction to generate electricity. *Electrolytic* cells are just the opposite: electricity from an external power source is used to drive a nonspontaneous redox reaction. An **external power source** needs to be added to the list of equipment in order to build an electrolytic cell.

82. The answer to this question is directly tied to two critical pieces of information about uranium emissions contained in the reading passage. First, the uranium compound emitted *high-energy radiation*. Furthermore, the emissions that exposed the photographic plate were *not* deflected by the charged plates. From these clues, we can conclude that **gamma radiation** is emitted by the uranium compound, since gamma radiation has the highest energy on the electromagnetic spectrum and, since it is pure energy, has neither mass nor charge.

83. When a uranium atom emits an alpha particle, it loses two protons and two neutrons from its nucleus, but most of the mass of the nucleus is left behind in the "daughter nuclide." To balance a nuclear equation, the sum of the atomic numbers (bottom) and the mass numbers (top) must be equal on both sides of the arrow. The nuclide that correctly completes the uranium alpha decay equation is

$$^{234}_{90}\text{Th}$$

84. Rewriting U-233 in isotopic notation is helpful for finding the number of neutrons contained in its nucleus. The atomic number of uranium is 92, which means that there are 92 protons contained in the nucleus of every uranium atom.

$$^{233}_{92}\text{U}$$

Subtract the atomic number (the bottom number) from the mass number (the top number) to find that there are **141 neutrons** in an atom of U-233.

85. The spontaneous emission of an alpha or a beta particle happens during the process of **radioactive decay**, which is also called **natural transmutation**.

Mark (✓) the questions you answered correctly. Count the number of checks and follow the formulas given to determine your score on each topic.

Core Area	☐ Questions Answered Correctly

69: (1)

Section M—Math Skills
☐ Number of checks ÷ 1 × 100 = ___ %

1, 2, 29, 31, 32, 72–74, 84: (9)

Section I—Atomic Concepts
☐ Number of checks ÷ 9 × 100 = ___ %

3–5, 12, 33, 36, 54: (7)

Section II—Periodic Table
☐ Number of checks ÷ 7 × 100 = ___ %

6, 34, 35, 37, 38, 52, 59, 60, 70, 75: (10)

Section III—Moles/Stoichiometry
☐ Number of checks ÷ 10 × 100 = ___ %

9, 39, 41, 51, 66–68: (7)

Section IV—Chemical Bonding
☐ Number of checks ÷ 7 × 100 = ___ %

7, 8, 11, 14, 15, 17, 19, 42, 43, 48, 53, 55–58, 63–65, 71, 76: (20)

Section V—Physical Behavior of Matter
☐ Number of checks ÷ 20 × 100 = ___ %

13, 16, 18, 20, 40, 44: (6)

Section VI—Kinetics and Equilibrium
☐ Number of checks ÷ 6 × 100 = ___ %

21–23, 45, 47, 50, 77: (7)

Section VII—Organic Chemistry
☐ Number of checks ÷ 7 × 100 = ___ %

Core Area	☐ Questions Answered Correctly

10, 24, 25, 78–81: (7)

Section VIII—Oxidation–Reduction
☐ Number of checks ÷ 7 × 100 = ___ %

6, 26, 27, 46, 49, 61, 62: (7)

Section IX—Acids, Bases, and Salts
☐ Number of checks ÷ 7 × 100 = ___ %

28, 30, 82, 83, 85: (5)

Section X—Nuclear Chemistry
☐ Number of checks ÷ 5 × 100 = ___ %

Examination August 2016

Chemistry—Physical Setting

PART A

Answer all questions in this part.

Directions (1–30): For *each* statement or question, write in the answer space the *number* of the word or expression that, of those given, best completes the statement or answers the question. Some questions may require the use of the *2011 Edition Reference Tables for Physical Setting/Chemistry*.

1 Which change occurs when an atom in an excited state returns to the ground state?

(1) Energy is emitted.

(2) Energy is absorbed.

(3) The number of electrons decreases.

(4) The number of electrons increases.

1 _____

2 The valence electrons in an atom of phosphorus in the ground state are all found in

(1) the first shell   (3) the third shell

(2) the second shell   (4) the fourth shell

2 _____

3 Which two elements have the most similar chemical properties?

(1) beryllium and magnesium

(2) hydrogen and helium

(3) phosphorus and sulfur

(4) potassium and strontium

3 _____

4 Which phrase describes a compound that consists of two elements?

   (1) a mixture in which the elements are in a variable proportion

   (2) a mixture in which the elements are in a fixed proportion

   (3) a substance in which the elements are chemically combined in a variable proportion

   (4) a substance in which the elements are chemically combined in a fixed proportion   4 _____

5 The formula mass of a compound is the

   (1) sum of the atomic masses of its atoms

   (2) sum of the atomic numbers of its atoms

   (3) product of the atomic masses of its atoms

   (4) product of the atomic numbers of its atoms   5 _____

6 The arrangement of the elements from left to right in Period 4 on the Periodic Table is based on

   (1) atomic mass

   (2) atomic number

   (3) the number of electron shells

   (4) the number of oxidation states   6 _____

7 Which diatomic molecule is formed when the two atoms share six electrons?

   (1) H_2 (3) O_2

   (2) N_2 (4) F_2 7 _____

8 Which formula represents a polar molecule?

   (1) O_2 (3) NH_3

   (2) CO_2 (4) CH_4 8 _____

9 Which element is *least* likely to undergo a chemical reaction?

   (1) lithium   (3) fluorine

   (2) carbon   (4) neon   9 _____

10 Which element has a melting point higher than the melting point of rhenium?

(1) iridium   (3) tantalum
(2) osmium   (4) tungsten   10 _____

11 Which property can be defined as the ability of a substance to be hammered into thin sheets?

(1) conductivity   (3) melting point
(2) malleability   (4) solubility   11 _____

12 Which list of elements consists of a metal, a metalloid, and a noble gas?

(1) aluminum, sulfur, argon
(2) magnesium, sodium, sulfur
(3) sodium, silicon, argon
(4) silicon, phosphorus, chlorine   12 _____

13 Which sample of matter has a crystal structure?

(1) $Hg(\ell)$ (3) $NaCl(s)$
(2) $H_2O(\ell)$ (4) $CH_4(g)$ 13 _____

14 One mole of liquid water and one mole of solid water have *different*

(1) masses   (3) empirical formulas
(2) properties   (4) gram-formula masses   14 _____

15 Which substance can *not* be broken down by a chemical change?

(1) butanal   (3) gold
(2) propene   (4) water   15 _____

16 Which statement describes particles of an ideal gas, based on the kinetic molecular theory?

(1) Gas particles are separated by distances smaller than the size of the gas particles.
(2) Gas particles do not transfer energy to each other when they collide.
(3) Gas particles have no attractive forces between them.
(4) Gas particles move in predictable, circular motion.   16 _____

17 Which expression could represent the concentration of a solution?

    (1)   3.5 g   (3)   3.5 mL

    (2)   3.5 M   (4)   3.5 mol   17 _____

18 Which form of energy is associated with the random motion of the particles in a sample of water?

    (1)   chemical energy   (3)   nuclear energy

    (2)   electrical energy   (4)   thermal energy   18 _____

19 Which change is most likely to occur when a molecule of H_2 and a molecule of I_2 collide with proper orientation and sufficient energy?

    (1)   a chemical change, because a compound is formed

    (2)   a chemical change, because an element is formed

    (3)   a physical change, because a compound is formed

    (4)   a physical change, because an element is formed   19 _____

20 Which changes can reach dynamic equilibrium?

    (1)   nuclear changes, only

    (2)   chemical changes, only

    (3)   nuclear and physical changes

    (4)   chemical and physical changes   20 _____

21 What occurs when a reaction reaches equilibrium?

    (1)   The concentration of the reactants increases.

    (2)   The concentration of the products increases.

    (3)   The rate of the forward reaction is equal to the rate of the reverse reaction.

    (4)   The rate of the forward reaction is slower than the rate of the reverse reaction.   21 _____

22 In terms of potential energy, PE, which expression defines the heat of reaction for a chemical change?

    (1) $PE_{products} - PE_{reactants}$ (3) $\dfrac{PE_{products}}{PE_{reactants}}$

    (2) $PE_{reactants} - PE_{products}$ (4) $\dfrac{PE_{reactants}}{PE_{products}}$ 22 _____

23 Systems in nature tend to undergo changes that result in

  (1) lower energy and lower entropy
  (2) lower energy and higher entropy
  (3) higher energy and lower entropy
  (4) higher energy and higher entropy   23 _____

24 What occurs when Cr^{3+} ions are reduced to Cr^{2+} ions?

  (1) Electrons are lost and the oxidation number of chromium increases.
  (2) Electrons are lost and the oxidation number of chromium decreases.
  (3) Electrons are gained and the oxidation number of chromium increases.
  (4) Electrons are gained and the oxidation number of chromium decreases.   24 _____

25 Where do reduction and oxidation occur in an electrolytic cell?

  (1) Both occur at the anode.
  (2) Both occur at the cathode.
  (3) Reduction occurs at the anode, and oxidation occurs at the cathode.
  (4) Reduction occurs at the cathode, and oxidation occurs at the anode.   25 _____

26 Which compound is an electrolyte?

  (1) H_2O (3) H_3PO_4
  (2) C_2H_6 (4) CH_3OH 26 _____

27 When the hydronium ion concentration of an aqueous solution is increased by a factor of 10, the pH value of the solution

  (1) decreases by 1   (3) decreases by 10
  (2) increases by 1   (4) increases by 10   27 _____

28 The stability of isotopes is related to the ratio of which particles in the atoms?

  (1) electrons and protons   (3) neutrons and protons
  (2) electrons and positrons   (4) neutrons and positrons   28 _____

29 Which radioisotope has the fastest rate of decay?

(1) ^{14}C (3) ^{53}Fe

(2) ^{37}Ca (4) ^{42}K 29 _____

30 The atomic mass of an element is the weighted average of the atomic masses of

(1) the least abundant isotopes of the element
(2) the naturally occurring isotopes of the element
(3) the artificially produced isotopes of the element
(4) the natural and artificial isotopes of the element   30 _____

PART B–1

Answer all questions in this part.

Directions (31–50): For *each* statement or question, write in the answer space the *number* of the word or expression that, of those given, best completes the statement or answers the question. Some questions may require the use of the *2011 Edition Reference Tables for Physical Setting/Chemistry.*

31 Which list of elements is arranged in order of increasing electro-negativity?

(1) Be, Mg, Ca   (3) K, Ca, Sc

(2) F, Cl, Br   (4) Li, Na, K   31 _____

32 The table below gives the masses of two different subatomic particles found in an atom.

Subatomic Particles and Their Masses

Subatomic Particle	Mass (g)
X	1.67×10^{-24}
Z	9.11×10^{-28}

Which of the subatomic particles are each paired with their corresponding name?

(1) X, proton and Z, electron

(2) X, proton and Z, neutron

(3) X, neutron and Z, proton

(4) X, electron and Z, proton   32 _____

33 Which electron configuration represents an excited state for an atom of calcium?

(1) 2-8-7-1   (3) 2-8-7-3

(2) 2-8-7-2   (4) 2-8-8-2   33 _____

34 At STP, graphite and diamond are two solid forms of carbon. Which statement explains why these two forms of carbon differ in hardness?

(1) Graphite and diamond have different ionic radii.
(2) Graphite and diamond have different molecular structures.
(3) Graphite is a metal, but diamond is a nonmetal.
(4) Graphite is a good conductor of electricity, but diamond is a poor conductor of electricity.   34 _____

35 Which equation shows conservation of charge?

(1) $Cu + Ag^+ \rightarrow Cu^{2+} + Ag$
(2) $Mg + Zn^{2+} \rightarrow 2Mg^{2+} + Zn$
(3) $2F_2 + Br^- \rightarrow 2F^- + Br_2$
(4) $2I^- + Cl_2 \rightarrow I_2 + 2Cl^-$ 35 _____

36 What occurs when potassium reacts with chlorine to form potassium chloride?

(1) Electrons are shared and the bonding is ionic.
(2) Electrons are shared and the bonding is covalent.
(3) Electrons are transferred and the bonding is ionic.
(4) Electrons are transferred and the bonding is covalent.   36 _____

37 Given the balanced equation representing a reaction:

$$H_2 + energy \rightarrow H + H$$

What occurs as bonds are broken in one mole of H_2 molecules during this reaction?

(1) Energy is absorbed and one mole of unbonded hydrogen atoms is produced.
(2) Energy is absorbed and two moles of unbonded hydrogen atoms are produced.
(3) Energy is released and one mole of unbonded hydrogen atoms is produced.
(4) Energy is released and two moles of unbonded hydrogen atoms are produced.   37 _____

38 Which pair of atoms has the most polar bond?

(1) H−Br   (3) I−Br

(2) H−Cl   (4) I−Cl   38 _____

39 Which two notations represent isotopes of the same element?

(1) $^{14}_{7}N$ and $^{18}_{7}N$ (3) $^{14}_{7}N$ and $^{17}_{10}Ne$

(2) $^{20}_{7}N$ and $^{20}_{10}Ne$ (4) $^{19}_{7}N$ and $^{16}_{10}Ne$ 39 _____

40 The graph below shows the volume and the mass of four different substances at STP.

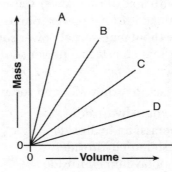

Which of the four substances has the *lowest* density?

(1) A   (3) C

(2) B   (4) D   40 _____

41 What is the total amount of heat required to completely melt 347 grams of ice at its melting point?

(1) 334 J   (3) 116 000 J

(2) 1450 J   (4) 784 000 J   41 _____

42 As the temperature of a reaction increases, it is expected that the reacting particles collide

(1) more often and with greater force

(2) more often and with less force

(3) less often and with greater force

(4) less often and with less force   42 _____

43 Given the formula representing a compound:

$$H-\overset{\overset{\displaystyle H}{|}}{\underset{\underset{\displaystyle H}{|}}{C}}-\overset{\overset{\displaystyle H}{|}}{\underset{\underset{\displaystyle H}{|}}{C}}-\overset{\overset{\displaystyle H}{|}}{\underset{\underset{\displaystyle H}{|}}{C}}-\overset{\overset{\displaystyle O}{\|}}{C}-\overset{\overset{\displaystyle H}{|}}{\underset{\underset{\displaystyle H}{|}}{C}}-\overset{\overset{\displaystyle H}{|}}{\underset{\underset{\displaystyle H}{|}}{C}}-H$$

What is an IUPAC name for this compound?

(1) ethyl propanoate   (3) 3-hexanone
(2) propyl ethanoate   (4) 4-hexanone   43 _____

44 A voltaic cell converts chemical energy to

(1) electrical energy with an external power source
(2) nuclear energy with an external power source
(3) electrical energy without an external power source
(4) nuclear energy without an external power source   44 _____

45 Which acid and base react to form water and sodium sulfate?

(1) sulfuric acid and sodium hydroxide
(2) sulfuric acid and potassium hydroxide
(3) sulfurous acid and sodium hydroxide
(4) sulfurous acid and potassium hydroxide   45 _____

46 Given the equation representing a reaction:

$$H_2CO_3 + NH_3 \rightarrow NH_4^+ + HCO_3^-$$

According to one acid-base theory, the compound NH_3 acts as a base because it

(1) accepts a hydrogen ion
(2) donates a hydrogen ion
(3) accepts a hydroxide ion
(4) donates a hydroxide ion   46 _____

47 Which statement describes characteristics of a 0.01 M KOH(aq) solution?

  (1) The solution is acidic with a pH less than 7.
  (2) The solution is acidic with a pH greater than 7.
  (3) The solution is basic with a pH less than 7.
  (4) The solution is basic with a pH greater than 7.   47 _____

48 Four statements about the development of the atomic model are shown below.

  A: Electrons have wavelike properties.

  B: Atoms have small, negatively charged particles.

  C: The center of an atom is a small, dense nucleus.

  D: Atoms are hard, indivisible spheres.

Which order of statements represents the historical development of the atomic model?

  (1) $C \rightarrow D \rightarrow A \rightarrow B$ (3) $D \rightarrow B \rightarrow A \rightarrow C$
  (2) $C \rightarrow D \rightarrow B \rightarrow A$ (4) $D \rightarrow B \rightarrow C \rightarrow A$ 48 _____

49 Five cubes of iron are tested in a laboratory. The tests and the results are shown in the table below.

Iron Tests and the Results

Test	Procedure	Result
1	A cube of Fe is hit with a hammer.	The cube is flattened.
2	A cube of Fe is placed in 3 M HCl(aq).	Bubbles of gas form.
3	A cube of Fe is heated to 1811 K.	The cube melts.
4	A cube of Fe is left in damp air.	The cube rusts.
5	A cube of Fe is placed in water.	The cube sinks.

Which tests demonstrate chemical properties?

  (1) 1, 3, and 4   (3) 2 and 4
  (2) 1, 3, and 5   (4) 2 and 5   49 _____

50 A rigid cylinder with a movable piston contains a sample of helium gas. The temperature of the gas is held constant as the piston is pulled outward. Which graph represents the relationship between the volume of the gas and the pressure of the gas?

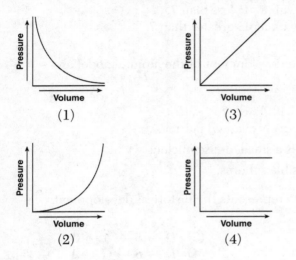

(1)   (3)

(2)   (4)

50 _____

PART B–2

Answer all questions in this part.

Directions (51–65): Record your answers on the answer sheet provided. Some questions may require the use of the *2011 Edition Reference Tables for Physical Setting/Chemistry*.

51 What is the empirical formula for C_6H_{12}? [1]

52 Using Table *G*, determine the minimum mass of NaCl that must be dissolved in 200. grams of water to produce a saturated solution at 90.°C. [1]

53 State the physical property that makes it possible to separate a solution by distillation. [1]

Base your answers to questions 54 and 55 on the information below and on your knowledge of chemistry.

A beaker contains a liquid sample of a molecular substance. Both the beaker and the liquid are at 194 K. The graph below represents the relationship between temperature and time as the beaker and its contents are cooled for 12 minutes in a refrigerated chamber.

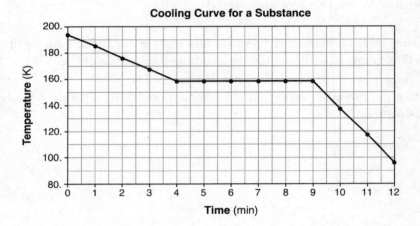

54 State what happens to the average kinetic energy of the molecules in the sample during the first 3 minutes.   [1]

55 Identify the physical change occurring during the time interval, minute 4 to minute 9.   [1]

Base your answers to questions 56 through 59 on the information below and on your knowledge of chemistry.

The equation below represents a reaction between propene and hydrogen bromide.

$$H-\underset{\underset{H}{|}}{\overset{\overset{H}{|}}{C}}-\underset{\underset{H}{|}}{\overset{\overset{H}{|}}{C}}=C-H \; + \; H-Br \longrightarrow H-\underset{\underset{H}{|}}{\overset{\overset{H}{|}}{C}}-\underset{\underset{Br}{|}}{\overset{\overset{H}{|}}{C}}-\underset{\underset{H}{|}}{\overset{\overset{H}{|}}{C}}-H$$

Cyclopropane, an isomer of propene, has a boiling point of −33°C at standard pressure and is represented by the formula below.

56 Explain why this reaction can be classified as a synthesis reaction.   [1]

57 Identify the class of organic compounds to which the product of this reaction belongs.   [1]

58 Explain, in terms of molecular formulas and structural formulas, why cyclopropane is an isomer of propene.   [1]

59 Convert the boiling point of cyclopropane at standard pressure to kelvins.   [1]

Base your answers to questions 60 through 63 on the information below and on your knowledge of chemistry.

The radius of a lithium atom is 130. picometers, and the radius of a fluorine atom is 60. picometers. The radius of a lithium ion, Li^+, is 59 picometers, and the radius of a fluoride ion, F^-, is 133 picometers.

60 Compare the radius of a fluoride ion to the radius of a fluorine atom.   [1]

61 Explain, in terms of subatomic particles, why the radius of a lithium ion is smaller than the radius of a lithium atom.   [1]

62 In the space *on your answer sheet*, draw a Lewis electron-dot diagram for a fluoride ion.   [1]

63 Describe the general trend in atomic radius as each element in Period 2 is considered in order from left to right.   [1]

Base your answers to questions 64 and 65 on the information below and on your knowledge of chemistry.

Nuclear fission reactions can produce different radioisotopes. One of these radioisotopes is Te-137, which has a half-life of 2.5 seconds. The diagram below represents one of the many nuclear fission reactions.

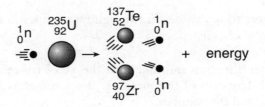

64 State evidence that this nuclear reaction represents transmutation.   [1]

65 Complete the nuclear equation *on your answer sheet* for the beta decay of Zr-97, by writing an isotopic notation for the missing product.   [1]

PART C

Answer all questions in this part.

Directions (66–85): Record your answers on the answer sheet provided. Some questions may require the use of the *2011 Edition Reference Tables for Physical Setting/Chemistry*.

Base your answers to questions 66 through 69 on the information below and on your knowledge of chemistry.

Stamping an identification number into the steel frame of a bicycle compresses the crystal structure of the metal. If the number is filed off, there are scientific ways to reveal the number.

One method is to apply aqueous copper(II) chloride to the number area. The Cu^{2+} ions react with some iron atoms in the steel frame, producing copper atoms that show the pattern of the number. The ionic equation below represents this reaction.

$$Fe(s) + Cu^{2+}(aq) \rightarrow Fe^{2+}(aq) + Cu(s)$$

Another method is to apply hydrochloric acid to the number area. The acid reacts with the iron, producing bubbles of hydrogen gas. The bubbles form faster where the metal was compressed, so the number becomes visible. The equation below represents this reaction.

$$2HCl(aq) + Fe(s) \rightarrow FeCl_2(aq) + H_2(g)$$

66 Explain why the Fe atoms in the bicycle frame react with the Cu^{2+} ions.   [1]

67 Determine the number of moles of hydrogen gas produced when 0.001 mole of HCl(aq) reacts completely with the iron metal.   [1]

68 Write a balanced half-reaction equation for the reduction of the hydrogen ions to hydrogen gas.   [1]

69 Describe *one* change in the HCl(aq) that will increase the rate at which hydrogen bubbles are produced when the acid is applied to the steel frame.   [1]

Base your answers to questions 70 through 73 on the information below and on your knowledge of chemistry.

In an investigation, aqueous solutions are prepared by completely dissolving a different amount of NaCl(s) in each of four beakers containing 100.00 grams of $H_2O(\ell)$ at room temperature. Each solution is heated and the temperature at which boiling occurred is measured. The data are recorded in the table below.

Boiling Point Data for Four NaCl(aq) Solutions

Beaker Number	Mass of $H_2O(\ell)$ (g)	Mass of NaCl(s) Dissolved (g)	Boiling Point of Solution (°C)
1	100.00	8.76	101.5
2	100.00	17.52	103.1
3	100.00	26.28	104.6
4	100.00	35.04	106.1

70 Identify the solute and the solvent used in this investigation.   [1]

71 Show a numerical setup for calculating the percent by mass of NaCl in the solution in beaker 4.   [1]

72 Explain, in terms of ions, why the ability to conduct an electric current is greater for the solution in beaker 4 than for the solution in beaker 1.   [1]

73 State the relationship between the concentration of ions and the boiling point for these solutions.   [1]

Base your answers to questions 74 through 76 on the information below and on your knowledge of chemistry.

One type of voltaic cell, called a mercury battery, uses zinc and mercury(II) oxide to generate an electric current. Mercury batteries were used because of their miniature size, even though mercury is toxic. The overall reaction for a mercury battery is given in the equation below.

$$Zn(s) + HgO(s) \rightarrow ZnO(s) + Hg(\ell)$$

74 Determine the change in the oxidation number of zinc during the operation of the cell.   [1]

75 Compare the number of moles of electrons lost to the number of moles of electrons gained during the reaction.   [1]

76 Using information in the passage, state *one* risk and *one* benefit of using a mercury battery.   [1]

Base your answers to questions 77 through 80 on the information below and on your knowledge of chemistry.

A company produces a colorless vinegar that is 5.0% $HC_2H_3O_2$ in water. Using thymol blue as an indicator, a student titrates a 15.0-milliliter sample of the vinegar with 43.1 milliliters of a 0.30 M NaOH(aq) solution until the acid is neutralized.

77 Based on Table *M*, what is the color of the indicator in the vinegar solution before any base is added? [1]

78 Identify the negative ion in the NaOH(aq) used in this titration. [1]

79 The concentration of the base used in this titration is expressed to what number of significant figures? [1]

80 Determine the molarity of the $HC_2H_3O_2$ in the vinegar sample, using the titration data. [1]

Base your answers to questions 81 through 85 on the information below and on your knowledge of chemistry.

In industry, ethanol is primarily produced by two different reactions. One process involves the reaction of glucose in the presence of an enzyme that acts as a catalyst. The equation below represents this reaction.

$$\text{Equation 1: } \underset{\text{glucose}}{C_6H_{12}O_6} \xrightarrow{\text{enzyme}} \underset{\text{ethanol}}{2CH_3CH_2OH} + 2CO_2$$

In another reaction, ethanol is produced from ethene and water. The equation below represents this reaction in which H_2SO_4 is a catalyst.

$$\text{Equation 2: } CH_2CH_2 + H_2O \xrightarrow{H_2SO_4} CH_3CH_2OH$$

Industrial ethanol can be oxidized using a catalyst to produce ethanal. The equation representing this oxidation is shown below.

$$\text{Equation 3: } CH_3CH_2OH \xrightarrow{\text{catalyst}} CH_3CHO + H_2$$

81 Identify the element that causes the reactant in equation 1 to be classified as an organic compound.   [1]

82 Identify the type of organic reaction represented by equation 1.   [1]

83 Explain why the hydrocarbon in equation 2 is unsaturated.   [1]

84 Explain, in terms of intermolecular forces, why ethanol has a much higher boiling point than ethene, at standard pressure.   [1]

85 Draw a structural formula for the organic product in equation 3.   [1]

Answer Sheet
August 2016

Chemistry—Physical Setting

PART B–2

51 _____

52 _____ g

53 _____

54 _____

_____

_____

55 _____

56 _____

_____

_____

57 _____

58 _____

_____

_____

59 _____ **K**

60 _____

_____

_____

61 _____

_____

_____

62

63 _____

_____

_____

64 _____

_____

_____

65 $^{97}_{40}\text{Zr} \rightarrow \ ^{0}_{-1}\text{e} +$ _____

PART C

66 _____

_____

_____

67 _____ **mol**

68 _____

69 _____

_____

_____

70 Solute:_____

Solvent:_____

71

72 _____

_____

_____

73 _____

_____

_____

74 From _____ to _____

75 _____

_____

_____

76 Risk: _____

Benefit: _____

77 _____

78 _____

79 _____

80 _____ **M**

81 _____

82 _____

83 _____

_____

_____

84 _____

_____

_____

85

Answers
August 2016
Chemistry—Physical Setting

Answer Key

PART A

1. 1	7. 2	13. 3	19. 1	25. 4
2. 3	8. 3	14. 2	20. 4	26. 3
3. 1	9. 4	15. 3	21. 3	27. 1
4. 4	10. 4	16. 3	22. 1	28. 3
5. 1	11. 2	17. 2	23. 2	29. 2
6. 2	12. 3	18. 4	24. 4	30. 2

PART B–1

31. 3	35. 4	39. 1	43. 3	47. 4
32. 1	36. 3	40. 4	44. 3	48. 4
33. 3	37. 2	41. 3	45. 1	49. 3
34. 2	38. 2	42. 1	46. 1	50. 1

PART B–2 and **PART C**. *See* **Answers Explained.**

Answers Explained

PART A

1. **1** An atom in an excited state contains at least one electron that has previously absorbed energy and has moved to a higher energy orbital within the atom. When the electron returns to the lowest energy position possible, also known as the ground state, the extra energy that was absorbed is emitted as a photon of electromagnetic energy.

2. **3** Valence electrons are the outermost electrons in an atom. The electron configuration for phosphorus, 2-8-5, is shown in the Periodic Table of the Elements in your Chemistry Reference Tables. This configuration shows that three principal energy levels (shells) within atoms of phosphorus contain electrons: 2 electrons in the first shell, 8 electrons in the second shell, and 5 electrons (the valence electrons) in the third shell.

3. **1** Elements in the same vertical group have the same number of valence electrons, which means that they have the same number of electrons available for bonding. Having the same number of valence electrons causes elements in the same group to have similar chemical properties.

4. **4** Compounds are substances composed of two or more elements that are chemically bonded in a fixed proportion. Compounds can be represented by chemical formulas (for example, H_2O), which show the relative proportion of each element in the compound.

WRONG CHOICES EXPLAINED:
(1), (2) A compound is a pure substance, whereas a mixture is made from two or more substances that are blended together but are not chemically bonded.

(3) As a pure substance, a compound is represented by a chemical formula that is always the same and does not vary. For example, water always has the same formula, H_2O.

5. **1** To find the formula mass (molar mass) of a compound, you must add up the atomic masses of all atoms in the compound. The atomic mass for each element is found on the Periodic Table of the Elements in your Reference Tables. For example, the formula mass of the compound carbon dioxide, CO_2, is found by adding the atomic mass of one carbon atom to the atomic mass of two oxygen atoms, as shown below:

$$\text{formula mass of } CO_2 =$$
$$[1 \times (\text{formula mass of C}) + 2 \times (\text{formula mass of oxygen})]$$

Substituting values for the formula masses of carbon and oxygen:

$$\text{formula mass of } CO_2 = [12.0 \text{ amu} + 2(16.0 \text{ amu})] = 44.0 \text{ amu}$$

6. **2** The Periodic Table of the Elements found in the Reference Tables illustrates that the elements are organized in order of increasing atomic number.

WRONG CHOICES EXPLAINED:
(1) The atomic mass *usually* increases from left to right across the Periodic Table of the Elements, but there are exceptions. In Period 4, for example, the atomic mass *decreases* between Co (58.9332 amu) and Ni (58.693 amu).
(3) The number of electron shells is 4 for all Period 4 elements. Notice that there are four values present in all of the electron configurations for Period 4 elements.
(4) The oxidation states for each element are shown in the upper right-hand corner of the element symbol boxes on the Periodic Table of the Elements. Oxidation states vary inconsistently across Period 4.

7. **2** All of the answer choices are covalently-bonded, diatomic molecules. A single covalent bond consists of one shared pair of electrons. For 6 electrons to be shared, there would need to be *three* covalent bonds joining the two atoms. Only nitrogen, N_2, contains a triple covalent bond. It forms this bond because each nitrogen atom has only 5 valence electrons and needs 3 *additional electrons* to form a stable octet of 8 valence electrons.

WRONG CHOICES EXPLAINED:
(1) Each H atom has only 1 valence electron and can form only one covalent bond.
(3) Each O atom has 6 valence electrons and will form two covalent bonds to obtain the stable octet.
(4) Each F atom has 7 valence electrons and will form just one covalent bond to obtain the stable octet.

8. **3** Ammonia (NH_3) molecules are polar, meaning that ammonia molecules have an end with a partial positive charge and an end with a partial negative charge as shown in the diagram below.

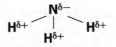

These partial charges arise from the polar covalent bonds between the nitrogen and hydrogen atoms. The nitrogen atom is more electronegative than the H atoms, so shared electrons will be attracted more strongly to nitrogen, giving it a partially negative charge.

WRONG CHOICES EXPLAINED:
(1), (2), (4) Each of these compounds contains polar covalent bonds in its molecules, but the overall symmetry of these molecules leads to a symmetrical charge distribution. A good expression to remember is that "symmetry cancels polarity."

9. **4** Neon is a noble gas in Group 18 of the Periodic Table of the Elements. Since elements in this group have 8 valence electrons, they are very stable and do not often form bonds with other elements.

10. **4** According to Reference Table S, the melting point of rhenium (Re) is 3458 K. Tungsten (W) has a melting point of 3695 K.

11. **2** This is a definition question. Malleability is a characteristic property of metals, along with excellent conductivity, ductility (the ability to be drawn into wires), and metallic luster.

12. **3** Metals are located to the left of the staircase dividing line on the Periodic Table of the Elements. Sodium, in Group 1, is a metal. Nonmetals are located to the right of the staircase dividing line on the Periodic Table of the Elements. Argon, in Group 18, is a noble gas. The six metalloids must be memorized. They are B, Si, As, Te, Sb, and Ge. It is easier to remember which elements are metalloids by visually memorizing the physical pattern that they make on the Periodic Table of the Elements.

WRONG CHOICES EXPLAINED:
(1), (2) These answer choices do not include a metalloid.
(4) This answer choice does not include a metal.

13. **3** Solids have a characteristic crystalline lattice structure. Both gases and liquids take the shape of their container because their particles have the ability to move.

14. **2** Water is water, no matter what phase it happens to be in. Properties of liquids and solids are very different, however. For example, ice floats on water, a demonstration of the different densities of these two phases.

WRONG CHOICES EXPLAINED:
(1) One mole of water has a mass of 18.0 grams, no matter what phase it is in.
(3) The empirical formula of water is the same as its molecular formula (H_2O). This formula is the same for all phases of water.
(4) The gram-formula (molar) mass of water is 18.0 g/mol, regardless of its phase.

15. **3** Chemical compounds can be broken down into their constituent elements by chemical change. Elements cannot be broken down by chemical change.

16. **3** One of the principles of the kinetic molecular theory of gases is that gas particles have no attraction for or repulsion to each other. This allows gas particles to have random motion and to expand to fill the space that they occupy. Other principles of the kinetic molecular theory of gases state that an ideal gas consists of particles that:

- Have mass
- Are separated by distances much greater than their size
- Move in random, constant, straight-line motion
- Undergo perfectly elastic collisions in which the total energy of the system remains the same before and after collisions

17. **2** All concentration units give information about how crowded solute particles are in a certain volume of solution. Common solution concentration units include molarity (M), which means moles per liter, percent by volume, percent by mass, and parts per million (ppm).

18. **4** Thermal energy is associated with the random motion of atoms and molecules. Thermal energy is created by the motion of atoms and molecules within an object. The faster the particles move, the more thermal energy the object has and the hotter the object will be. Chemical energy, on the other hand, refers to energy stored in chemical bonds.

19. **1** Proper orientation and sufficient energy are the two requirements for an effective collision. Effective collisions result in chemical reactions (chemical changes) in which chemical bonds are broken and formed. A chemical reaction between H_2 and I_2 would produce the compound HI.

20. **4** Some chemical and physical changes can reach the state of dynamic equilibrium, in which a reversible process occurs in both directions at equal rates. Chemical equilibrium occurs in a closed system when the forward and reverse chemical reactions occur at the same rate. An example of dynamic equilibrium in a physical change is solution equilibrium, when the rates of dissolving and precipitation in a saturated solution are equal.

WRONG CHOICES EXPLAINED:
(1), (3) One reason why nuclear changes can never reach a state of dynamic equilibrium is that the system cannot be closed. Nuclear particles and high-energy electromagnetic radiation that is released during nuclear reactions can travel through the walls of most containers.

(2) Some physical changes can also reach a state of dynamic equilibrium, as in the solution equilibrium described above. Vapor pressure equilibrium is another example of physical equilibrium, in which evaporation and condensation proceed at equal rates in a closed flask containing a pure liquid.

21. **3** When a chemical reaction reaches equilibrium, the rates of the forward and reverse reactions are equal. Since reactant and product molecules are forming at equal rates, the concentrations of all molecules remain constant at equilibrium.

22. **1** The heat of reaction (ΔH) is defined as the potential energy of the products of a chemical reaction minus the potential energy of the reactants. In other words, the heat of reaction is the net change in energy during the reaction. A good analogy for this is elevation change. If you walk up a hill, your elevation increases, which corresponds to a positive change in elevation (elevation at the end of your walk – elevation at the beginning). Likewise, walking down a hill would result in a negative change in elevation.

23. **2** Systems in nature tend to undergo changes that produce lower total energy and greater disorder or randomness (entropy). For example, an avalanche of rock debris down a hillside moves toward a position of lower potential energy and much greater randomness.

24. **4** The chemical process of reduction involves the gain of electrons. This is a definition you must know. A helpful mnemonic device for this definition is OIL RIG, which translates to <u>O</u>xidation <u>I</u>s <u>L</u>oss and <u>R</u>eduction <u>I</u>s <u>G</u>ain. If the Cr^{3+} ion gains 1 electron, the oxidation number of the ion decreases from 3+ to 2+.

25. **4** The anode is defined as the site of oxidation, and the cathode is defined as the site of reduction in all electrochemical cells, both voltaic and electrolytic. A great mnemonic expression for remembering this definition is the expression "An Ox, Red Cat" (oxidation occurs at the anode and reduction happens at the cathode).

26. **3** Electrolytes are compounds that dissociate into ions when dissolved in water, creating aqueous solutions that are capable of conducting electricity. Acids, bases, and ionic compounds in general are three classes of compounds that have electrolytic properties. H_3PO_4 is an acid that delivers H^+ ions into solution (see Reference Table *K*), so it is an electrolyte.

27. **1** On the pH scale, each decrease of one unit of pH represents a 10-fold increase in the hydronium ion concentration. This is because the pH scale relates to the base 10 exponent of the hydronium ion concentration.

28. **3** Neutrons and protons reside in the nucleus of an atom. The ratio between the two causes a nucleus to be either stable or unstable.

WRONG CHOICES EXPLAINED:
(1), (2) Electrons are found outside of the nucleus and therefore do not affect its stability.
(2), (4) Positrons are positively charged bits of antimatter. They are the size of an electron but have the opposite charge. Positrons are sometimes emitted by unstable nuclei as they undergo radioactive decay. They have nothing at all to do with the stability of an atom's nucleus.

29. **2** Look at Reference Table *N* for the half-lives of these radioisotopes. ^{37}Ca, with a half-life of only 182 ms, decays far faster than any of the other isotopes offered as answer choices.

30. **2** The average atomic mass of any element is calculated by taking the weighted average of all of the naturally occurring isotopes of that element. The percent abundance of each naturally occurring isotope is multiplied by its atomic mass, resulting in an average that reflects how abundant each isotope is in nature.

PART B–1

31. **3** Electronegativity data for many elements is listed in Reference Table S. Looking at data for the sets of elements shown as answer choices, only the progression K, Ca, Sc is arranged in order of increasing electronegativity. Electronegativity (attraction for electrons) increases from left to right across the Periodic Table of the Elements because of increasing nuclear charge.

32. **1** Consult Reference Table O for information about nuclear particles. According to this table, protons have a mass of 1 amu, which is actually equal to 1.67×10^{-24} grams. Electrons, which are identical to beta particles, have a mass of 0 according to Reference Table O. Of course, electrons do not really have a mass of 0, but they do have a mass that is far less than the mass of a proton.

33. **3** The ground-state electron configuration for calcium, 2-8-8-2, is found on the Periodic Table of the Elements in the Reference Tables. A calcium atom would be in an excited state if its configuration were 2-8-7-3 because this configuration shows that a third shell electron has moved to the fourth shell.

WRONG CHOICES EXPLAINED:
(1), (2) Calcium atoms have 20 electrons. Neither of these electron configurations has the correct number of electrons.
(4) This is the ground-state electron configuration for calcium. All of the electrons are in the lowest possible energy levels.

34. **2** Diamond and graphite are two *allotropes* of carbon, meaning that they are naturally occurring forms of carbon that have different structural arrangements. Diamond is arranged in a three-dimensional tetrahedral structure, and graphite is arranged in flat sheets, as shown in the illustrations below. Having different structures causes these substances to have different properties.

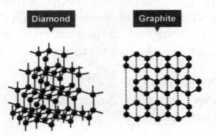

Diamond   Graphite

35. **4** In a redox reaction, the number of electrons lost is equal to the number of electrons gained. In the following equation:

$$2I^- + Cl_2 \rightarrow I_2 + 2Cl^-$$

1 electron is lost by each I^- ion as it oxidizes to form I_2, and 1 electron is gained by each Cl atom as it reduces to form Cl^-.

WRONG CHOICES EXPLAINED:
(1) Cu would have to lose 2 electrons to form Cu^{2+}, while Ag^+ would only have to gain 1 electron to form Ag^0.
(2), (3) These equations are not balanced for mass, let alone charge!

36. **3** The key to answering this question is to recognize that electrons are transferred from one atom to another only when an ionic bond forms (electrons are shared in covalent bonds). Ionic bonds form when metals, such as potassium, bond with nonmetals, such as chlorine.

37. **2** Energy is absorbed when bonds break, and energy is released or given off when bonds form. The equation in this question shows a bond breaking as H_2 absorbs energy and breaks apart into two separate atoms.

38. **2** Polarity across a bond is caused by the difference in electronegativity between the two elements involved in the bond. If the electronegativity difference is small, the bond is less polar, whereas if the electronegativity difference is large, the bond will be highly polar with a strongly negative end and a strongly positive end. Look up the electronegativity of the elements in this question on Reference Table *S* to confirm that the H–Cl bond has a higher electronegativity difference than any of the other answer choices.

39. **1** Isotopes are different versions of an element with different numbers of neutrons. All nitrogen atoms must have an atomic number (the bottom number) of 7. Different isotopes of nitrogen will have different mass numbers (the top number), reflecting the number of neutrons they contain.

WRONG CHOICES EXPLAINED:
(2), (3), (4) Nitrogen (N) and neon (Ne) are different elements. Isotopes are different versions of *the same element*.

40. **4** Density equals mass divided by volume. This formula is found on Reference Table T. Since mass is the y-axis coordinate, and volume is the x-axis coordinate, the slope of each line, which is equal to $\dfrac{\Delta y}{\Delta x}$, is equal to the density of the substance. The graph for substance D has the lowest slope and therefore the lowest density.

41. **3** The last page of the Reference Tables, Reference Table T, lists the formula to find the heat required to melt any pure substance in the solid phase:

$$q = mH_f$$

where q = heat, m = mass, and H_f is the heat of fusion. The mass is given as 347 grams, and the heat of fusion for ice (water), found on Reference Table B, is 334 J/g. Plug these values into the formula and solve:

$$q = (347\ \text{g})(334\ \text{J/g}) \approx 116\ 000\ \text{J}$$

42. **1** Temperature is directly related to the average kinetic energy of particles in any sample of matter. This relationship is one of the key points in the kinetic molecular theory of matter. Kinetic energy is the energy of motion. As the temperature of a reaction increases, the reacting particles will move faster and will collide more often and with more force.

43. **3** The compound in this question contains the functional group of a ketone, as shown on Reference Table R:

$$\overset{\displaystyle O}{\underset{\displaystyle -\,C\,-}{\|}}$$

From the example for ketones in this table, we know that the suffix for ketones is "-one." There are 6 carbon atoms in the compound in this question, so the name will have the prefix "hex-." Additionally, a number is assigned for the position of the functional group. IUPAC rules for naming organic compounds state that the carbon atoms should be numbered so as to give the functional group the lowest possible numbered position. The functional group is on the third carbon from the right, so putting all of the pieces together, the correct name for this compound is 3-hexanone.

WRONG CHOICES EXPLAINED:

(1), (2) These names are for esters, not ketones.

(4) While the ketone functional group is on the fourth carbon atom from the left, IUPAC rules require us to assign the lowest possible number to the position of the functional group. We can place the ketone group on the third carbon if we number from the right.

44. **3** A voltaic cell spontaneously converts chemical energy to electrical energy. No outside energy source is required.

45. **1** The neutralization reaction between sulfuric acid and sodium hydroxide is

$$H_2SO_4 + 2NaOH \rightarrow Na_2SO_4 + 2H_2O$$

Formulas for both sulfuric acid and sodium hydroxide can be found on Reference Tables K and L, respectively. Sodium sulfate, Na_2SO_4, is a product of this neutralization.

46. **1** The Brønsted–Lowry theory of acids and bases defines a base as a proton (H^+) acceptor. In the equation shown, NH_3 accepts (gains) a H^+ ion to form the ammonium ion, NH_4^+.

47. **4** KOH, potassium hydroxide, is listed on Reference Table L as a common base. Basic solutions have pH values greater than 7, which is the pH of neutral water.

48. **4** The idea of the atom as a hard, indivisible sphere originated in ancient Greece. Electrons were discovered in the late 19th century in J.J. Thomson's famous cathode ray experiment, and, shortly after that, the nucleus was discovered in Ernest Rutherford's gold foil experiment. Most recently, in the quantum model of the atom, electrons have been demonstrated to have properties of both waves and particles, a phenomenon known as wave-particle duality.

49. **3** A chemical property involves how an element or compound will react to form a new substance. Examples of chemical properties include the ability to oxidize (as in the process of rusting) and the ability to react with acid. Physical properties are properties that can be observed without changing the chemical identity of a substance.

50. **1** Pressure and volume are inversely related. This relationship is known as Boyle's Law:

$$P \times V = k$$

where k is a constant. Since the product of pressure and volume is a constant, if one of them increases, the other must decrease. The inverse nature of this relationship is shown in the graph for choice (1).

PART B–2

[All questions in Part B–2 are worth 1 point.]

51. Empirical formulas give the smallest whole-number ratio of atoms in a compound. An empirical formula reduces the subscripts in a compound formula into "lowest terms." The empirical formula for C_6H_{12} is **CH$_2$**.

52. Reading Table G in the Reference Tables, it can be seen that 40 grams of NaCl dissolved in 100 grams of water produces a saturated solution. To saturate 200 grams of water, **80 grams of NaCl** is the minimum amount that must be dissolved.

(NOTE: Credit was awarded for all answers in the range of **78–82 grams of NaCl**.)

53. Distillation is used to separate the components of a liquid mixture based on their different **boiling points**.

54. Temperature is directly related to kinetic energy, which is the energy of motion of particles in any sample of matter. According to the graph, the temperature of the substance decreases during the first 3 minutes. Due to the direct relationship between temperature and kinetic energy, the **kinetic energy of the molecules in the sample decreases** during the first 3 minutes.

55. The temperature does not change between minutes 4 and 9, which means that the kinetic energy of the sample is constant. Since the sample is still being cooled, energy is leaving the sample in the form of *potential energy*. As potential energy decreases, particles in the liquid sample move closer together, and a phase change to the solid state takes place. This process is known as **freezing**, or **solidification**.

56. Synthesis reactions follow the pattern A + B → AB. In this case, **two reactants combine to form one product**.

57. Use Reference Table R to classify the organic compound formed in this reaction. Since the compound formed contains a bromine (Br) atom, it is classified as a **halide (halocarbon)**.

58. **Isomers are organic compounds that have the same molecular formula but different structural formulas and properties.** In other words, compounds that are isomers of each other will consist of the same number of atoms of each element, but the atoms will be arranged differently in the different compounds. **Both propene and cyclopropane have the molecular formula, C_3H_6, but one compound has a double bond and the other has a ring structure.**
(NOTE: Both answers in bold are acceptable.)

59. The last formula on Reference Table T is the formula used to convert from the Celsius to the Kelvin scale and vice versa.

$$K = °C + 273$$

According to the information given in this problem, cyclopropane boils at $-33°C$.
Plugging this information into the formula:

$$K = -33°C + 273 = \textbf{240. K}$$

60. The answer to this question can be found directly in the information provided. The fluorine atom has a radius of 60. picometers, and the fluoride ion has a radius of 133 picometers. **The radius of the fluoride ion is larger than the radius of the fluorine atom.** The extra electron gained to form the F^{-1} ion is repelled by the 7 electrons already in fluorine's valence shell, and the added repulsion increases the size of the electron cloud.

61. This question states that students must answer "in terms of subatomic particles." In other words, students must include information about protons, neutrons, or electrons in their answer. Electrons are subatomic particles that atoms gain or lose as they form ions, so students must include information about electrons in their response. A simplified diagram of a lithium atom is shown below.

When a lithium atom becomes a lithium ion, Li^+, **the second-shell electron is lost, so the lithium ion only has electrons in the first shell**. Another acceptable answer for this question is simply to say that the **lithium ion has one less electron**.

62. Lewis electron-dot diagrams for ions are drawn to show the valence electrons in the ion *and* the charge of the ion. Fluorine, as an element, has 7 valence electrons, but the fluoride ion contains 1 additional electron, giving it a charge of -1, as shown in the correct diagrams below. The common convention is to draw brackets around the octet of valence electrons for nonmetal ions showing that these electrons have been *transferred completely* to the ion and will not be shared with another atom.

$[\overset{..}{\underset{..}{:F:}}]^-$ *or* $:\overset{..}{\underset{..}{F}}:^-$ are acceptable answers.

63. **The atomic radius decreases as elements are considered in order from left to right across Period 2 of the Periodic Table of the Elements.** This trend is caused by increasing nuclear charge, which draws the electron cloud in closer, thereby shrinking the atomic radius. Since atomic radius data is provided in Reference Table *S*, you can always look up the values and determine the trend.

64. The term "transmutation" refers to any change in the nucleus of an atom that converts it from one element to another. The diagram shows that **uranium is changing into different elements**.

65. When a zirconium-97 atom emits a beta particle, it loses a particle with the mass and charge of an electron from its nucleus. To balance a nuclear equation, the sum of the atomic numbers (bottom) and the mass numbers (top) must be equal on both sides of the arrow. The nuclide that correctly completes the zirconium-97 beta decay equation is

$$^{97}_{41}\text{Nb}$$

(Note: Other acceptable forms of this answer are Nb-97 and niobium-97.)

PART C

[All questions in Part C are worth 1 point.]

66. Reference Table *J* lists metals from the most active metals at the top to the least active metals at the bottom. A more active metal will oxidize in the presence of the ion of a less active metal. Since iron is higher than copper on Table *J*, **iron is a more active metal than copper**.

67. All coefficients in chemical reactions represent particles or moles. Mole ratios can be used to find the number of moles of any substance in a chemical equation if the moles of another substance are known. Use the second equation in the information listed, which shows H_2 gas forming from the reaction between Fe and HCl. Two methods of setting up this problem are:

$$\frac{.001 \text{ mol HCl(aq)}}{1} \times \frac{1 \text{ mol } H_2(g)}{2 \text{ mol HCl(aq)}} = \textbf{0.0005 mol } \textbf{H}_2(\textbf{g})$$

OR

Set up mole ratios:

$$\frac{1 \text{ mol } H_2(g)}{2 \text{ mol HCl(aq)}} = \frac{x \text{ mol } H_2(g)}{0.001 \text{ mol HCl(aq)}}$$

$$x = \textbf{0.0005 mol } \textbf{H}_2(\textbf{g})$$

68. The second equation shows that the H^+ ions from hydrochloric acid react with solid iron to form hydrogen gas, H_2. Pure elements have an oxidation number of 0, so each H^+ ion that reacts must gain 1 electron to become a hydrogen atom. This reduction (gain of electrons) is shown in the balanced half-reaction below:

$$\textbf{2H}^+(\textbf{aq}) + \textbf{2e}^- \rightarrow \textbf{H}_2{}^0(\textbf{g})$$

(NOTE: Showing the oxidation number on H_2 in the equation above is optional.)

69. Hydrogen bubbles will be produced more quickly if the rate of the chemical reaction between hydrochloric acid and iron increases. The rate of a chemical reaction increases when the number of collisions between reactant particles increases. **Increasing the concentration of the HCl(aq)** would cause there to be more HCl particles available to collide with Fe(s). Another approach would be to **increase the temperature of the HCl(aq)**. The HCl molecules would move faster at the higher temperature and would collide more often with Fe(s).

70. The solute in a solution is the substance that dissolves, and the solvent is the substance into which the solute dissolves. **NaCl(s) is the solute, and $\textbf{H}_2\textbf{O}(\ell)$ is the solvent.**

71. Use the percent composition formula on Reference Table T:

$$\% \text{ composition by mass} = \frac{\text{mass of part}}{\text{mass of whole}} \times 100$$

In a solution, the "part" is the mass of the solute, and the "whole" is the mass of the *entire solution*. The solution in beaker 4 weighs a total of 135.04 g, which is the sum of 100.00 g $H_2O(l)$ plus 35.04 g of dissolved $NaCl(s)$. The correct numerical setup for this problem is therefore:

$$\textbf{\% composition by mass} = \frac{\textbf{35.04 g NaCl(s)}}{\textbf{135.04 g total}} \times \textbf{100}$$

(NOTE: This problem only asks for a numerical setup. You are NOT required to solve for the answer. If you DO solve for the actual percent composition and you solve incorrectly, you will LOSE the point for this question even if you have shown the correct numerical setup. If you wish to solve for the percent composition to check that your setup is correct, do so on scrap paper.)

72. Electrolytes are substances that dissolve in water to form aqueous solutions capable of conducting electricity. The more ions there are in a solution, the more electrically conductive the solution becomes. Since more $NaCl(s)$ was dissolved, **beaker 4 has more ions dissolved in aqueous solution than beaker 1**. The increased ion concentration causes beaker 4 to have a greater ability to conduct electricity.

73. **As the concentration of ions increases, the boiling point of the solution increases.** This relationship is supported by the data in the table provided. The mass of dissolved $NaCl(s)$ increases from beaker 1 through beaker 4, and the boiling point increases as well.

74. The $Zn(s)$ reactant is a pure element, and it has an oxidation number of 0. The compound $ZnO(s)$ that is produced in this reaction has an overall oxidation number of 0. Since oxygen has an oxidation number of -2, the zinc ion in this compound must have an oxidation number of $+2$, making the overall sum of the two oxidation numbers 0. Therefore, zinc's oxidation number changes from **0 to +2** as this cell operates.

75. Redox reactions are reactions that involve the transfer of electrons. The number of electrons lost by the substance that oxidizes must always be equal to

the number of electrons gained by the substance that reduces. In other words, **e$^-$ lost = e$^-$ gained**.

76. The answer to this question is stated directly in the information provided, making this a "careful reading" question. In the second sentence, we learn that **mercury is toxic (the risk), but its use made possible the production of miniature batteries (the benefit)**.

77. The information provided for this question states that vinegar contains acetic acid, $HC_2H_3O_2$. Acetic acid is listed as a common acid on Reference Table K. All acidic solutions have pH values less than 7, which is the pH for pure water. According to Reference Table M, thymol blue, the indicator used in this titration, is **yellow** in all solutions having a pH less than 8.0.

78. NaOH is a strong base that dissociates into ions when it is mixed with water to form an aqueous solution. The two ions produced when NaOH dissociates are Na^+ and **OH^-**.

79. The number 0.30 contains **two** significant figures.

80. Using the titration formula on Reference Table T:

$$M_AV_A = M_BV_B$$

in which M_A = the molarity of H^+, V_A = volume of acid, M_B = molarity of OH^-, and V_B = volume of base.
   Substituting the values from this problem:

$$M_A(15.0 \text{ mL}) = (0.30 \text{ M})(43.1 \text{ mL})$$
$$M_A = \frac{(0.30 \text{ M})(43.1 \text{ mL})}{(15.0 \text{ mL})} = \textbf{0.86 M}$$

81. All organic compounds contain **carbon** atoms.

82. Equation 1 shows glucose converting into ethanol in the presence of an enzyme catalyst. This is an example of a **fermentation** reaction.

83. Unsaturated hydrocarbons contain at least one carbon–carbon multiple bond. Memorize this definition! The compound CH_2CH_2 has the molecular formula C_2H_4, and, according to Reference Table Q, it must be an alkene since it

follows the general formula C_nH_{2n}. Alkenes have a double carbon–carbon bond, as illustrated in the structural formula for alkenes in Reference Table Q. Since **a double C=C bond is present**, this compound is an unsaturated hydrocarbon.

84. Ethanol has a much higher boiling point than the hydrocarbon ethene because ethanol molecules are attracted to each other by hydrogen bonding. Hydrogen bonding is an extremely strong type of intermolecular attraction. It occurs only in compounds that have one or more hydrogen atoms *directly bonded to* nitrogen, oxygen, or fluorine atoms, which are all highly electronegative. By this definition, it is clear that hydrogen bonding does not exist between ethene molecules. Since **ethanol has stronger intermolecular attractions than ethene**, molecules of ethanol have a harder time separating from each other, so boiling takes place at a higher temperature.

85. Reference Table R shows the structural formulas for major classes of organic compounds. The formula CH_3CHO follows the pattern shown for aldehydes. The aldehyde produced in equation 3 would have the structural formula shown below:

Mark (✓) the questions you answered correctly. Count the number of checks and follow the formulas given to determine your score on each topic.

Core Area	☐ Questions Answered Correctly

10, 40, 79: (3)

Section M—Math Skills
☐ Number of checks ÷ 3 × 100 = ___ %

1, 2, 30, 32, 33, 39, 48: (7)

Section I—Atomic Concepts
☐ Number of checks ÷ 7 × 100 = ___ %

3, 6, 11, 12, 31, 34, 63: (7)

Section II—Periodic Table
☐ Number of checks ÷ 7 × 100 = ___ %

4, 5, 15, 51, 56, 67: (6)

Section III—Moles/Stoichiometry
☐ Number of checks ÷ 6 × 100 = ___ %

7–9, 36–38, 60–62: (9)

Section IV—Chemical Bonding
☐ Number of checks ÷ 9 × 100 = ___ %

13–18, 41, 49, 50, 52–55, 59, 70, 71, 73, 84: (18)

Section V—Physical Behavior of Matter
☐ Number of checks ÷ 18 × 100 = ___ %

19, 20–23, 42, 69: (7)

Section VI—Kinetics and Equilibrium
☐ Number of checks ÷ 7 × 100 = ___ %

43, 57, 81–83, 85: (6)

Section VII—Organic Chemistry
☐ Number of checks ÷ 6 × 100 = ___ %

Core Area   ☐ Questions Answered Correctly

24, 25, 35, 44, 58, 66, 68, 74, 75: (9)

Section VIII—Oxidation–Reduction

☐ Number of checks ÷ 9 × 100 = ___ %

26, 27, 45–47, 72, 77, 78, 80: (9)

Section IX—Acids, Bases, and Salts

☐ Number of checks ÷ 9 × 100 = ___ %

28, 29, 64, 65: (4)

Section X—Nuclear Chemistry

☐ Number of checks ÷ 4 × 100 = ___ %

Examination
June 2017
Chemistry—Physical Setting

PART A

Answer all questions in this part.

Directions (1–30): For *each* statement or question, write in the answer space the *number* of the word or expression that, of those given, best completes the statement or answers the question. Some questions may require the use of the *2011 Edition Reference Tables for Physical Setting/Chemistry*.

1 Which statement describes the structure of an atom?

  (1) The nucleus contains positively charged electrons.
  (2) The nucleus contains negatively charged protons.
  (3) The nucleus has a positive charge and is surrounded by negatively charged electrons.
  (4) The nucleus has a negative charge and is surrounded by positively charged electrons.

      1 _____

Proton = + electron = −

2 Which term is defined as the region in an atom where an electron is most likely to be located?

  (1) nucleus   (3) quanta
  (2) orbital   (4) spectra

      2 _____

electrons orbit the atom

3 What is the number of electrons in an atom of scandium?

  (1) 21   (3) 45
  (2) 24   (4) 66

      3 _____

Atomic # = # of electrons

4 Which particle has the *least* mass?

(1) a proton   (3) a helium atom

(2) an electron   (4) a hydrogen atom   4 _____

electrons have almost no weight

5 Which electron transition in an excited atom results in a release of energy?

(1) first shell to the third shell

(2) second shell to the fourth shell

(3) third shell to the fourth shell

(4) fourth shell to the second shell   5 _____

release of energy decreases shell

6 On the Periodic Table, the number of protons in an atom of an element is indicated by its

(1) atomic mass

(2) atomic number

(3) selected oxidation states

(4) number of valence electrons   6 _____

atomic # = # of protons

7 Which type of formula shows an element symbol for each atom and a line for each bond between atoms?

(1) ionic   (3) empirical

(2) structural   (4) molecular   7 _____

8 What is conserved during all chemical reactions?

(1) charge   (3) vapor pressure

(2) density   (4) melting point   8 _____

9 In which type of reaction can two compounds exchange ions to form two different compounds?

(1) synthesis   (3) single replacement

(2) decomposition   (4) double replacement   9 _____

Two ions exchanging

10 At STP, two 5.0-gram solid samples of different ionic compounds have the same density. These solid samples could be differentiated by their

(1) mass   (3) temperature

(2) volume   (4) solubility in water   10 _____

Ionic compounds differ in solubility

11 What is the number of electrons shared between the atoms in an I_2 molecule?

(1) 7   (3) 8

(2) 2   (4) 4   11 _____

Both atoms need one more electron

12 Which substance has nonpolar covalent bonds?

(1) Cl_2 (3) SiO_2

(2) SO_3 (4) CCl_4 12 _____

Non polar covalent B between atoms w/same electronegativity

13 Compared to a potassium atom, a potassium ion has

(1) a smaller radius   (3) fewer protons

(2) a larger radius   (4) more protons   13 _____

Potassium ion loses a valence shell

14 Which form of energy is associated with the random motion of particles in a gas?

(1) chemical   (3) nuclear

(2) electrical   (4) thermal   14 _____

Greater heat = Greater speed

15 The average kinetic energy of water molecules *decreases* when

(1) $H_2O(\ell)$ at 337 K changes to $H_2O(\ell)$ at 300. K

(2) $H_2O(\ell)$ at 373 K changes to $H_2O(g)$ at 373 K

(3) $H_2O(s)$ at 200. K changes to $H_2O(s)$ at 237 K

(4) $H_2O(s)$ at 273 K changes to $H_2O(\ell)$ at 273 K   15 _____

Temp directly correlates to KE

16 The joule is a unit of

(1) concentration   (3) pressure

(2) energy   (4) volume   16 _____

17 Compared to a sample of helium at STP, the same sample of helium at a higher temperature and a lower pressure

(1) condenses to a liquid

(2) is more soluble in water

(3) forms diatomic molecules

(4) behaves more like an ideal gas   17 _____

18 A sample of a gas is in a sealed, rigid container that maintains a constant volume. Which changes occur between the gas particles when the sample is heated?

(1) The frequency of collisions increases, and the force of collisions decreases.

(2) The frequency of collisions increases, and the force of collisions increases.

(3) The frequency of collisions decreases, and the force of collisions decreases.

(4) The frequency of collisions decreases, and the force of collisions increases.

18 _____

Faster speed means stronger collision

19 At STP, which gaseous sample has the same number of molecules as 3.0 liters of $N_2(g)$?

(1) 6.0 L of $F_2(g)$ (3) 3.0 L of $H_2(g)$

(2) 4.5 L of $N_2(g)$ (4) 1.5 L of $Cl_2(g)$

19 _____

Number of particles corolates to volume

20 Distillation of crude oil from various parts of the world yields different percentages of hydrocarbons. Which statement explains these different percentages?

(1) Each component in a mixture has a different solubility in water.

(2) Hydrocarbons are organic compounds.

(3) The carbons in hydrocarbons may be bonded in chains or rings.

(4) The proportions of components in a mixture can vary.

20 _____

Every batch of crude oil will have different proportion

21 In which 1.0-gram sample are the particles arranged in a crystal structure?

(1) $CaCl_2(s)$ (3) $CH_3OH(\ell)$

(2) $C_2H_6(g)$ (4) $CaI_2(aq)$

21 _____

Solids tend to arrange in crystaline structures

22 When a reversible reaction is at equilibrium, the concentration of products and the concentration of reactants must be

(1) decreasing   (3) constant

(2) increasing   (4) equal

22 _____

Everything is constant at equilibrium

23 In chemical reactions, the difference between the potential energy of the products and the potential energy of the reactants is equal to the

(1) activation energy   (3) heat of reaction

(2) ionization energy   (4) heat of vaporization   23 _____

Energy can never be created or destroyed

24 What occurs when a catalyst is added to a chemical reaction?

(1) an alternate reaction pathway with a lower activation energy

(2) an alternate reaction pathway with a higher activation energy

(3) the same reaction pathway with a lower activation energy

(4) the same reaction pathway with a higher activation energy   24 _____

Catalysts reduce activation energy

25 What is the name of the compound with the formula $CH_3CH_2CH_2NH_2$?

(1) 1-propanol   (3) propanal

(2) 1-propanamine   (4) propanamide   25 _____

Reference table P and Q

26 Which compound is an isomer of $C_2H_5OC_2H_5$?

(1) CH_3COOH (3) $C_3H_7COCH_3$

(2) $C_2H_5COOCH_3$ (4) C_4H_9OH 26 _____

Same # of atoms arranged differently

27 Ethanoic acid and 1-butanol can react to produce water and a compound classified as an

(1) aldehyde   (3) ester

(2) amide   (4) ether   27 _____

Just draw the molecule

28 During an oxidation-reduction reaction, the number of electrons gained is

(1) equal to the number of electrons lost

(2) equal to the number of protons gained

(3) less than the number of electrons lost

(4) less than the number of protons gained   28 _____

Matter cannot be created or destroyed

29 Which process requires energy for a nonspontaneous redox reaction to occur?

(1) deposition   (3) alpha decay
(2) electrolysis   (4) chromatography   29 ____

electrolysis needs electricity

30 Which pair of compounds represents one Arrhenius acid and one Arrhenius base?

(1) CH_3OH and NaOH   (3) HNO_3 and NaOH
(2) CH_3OH and HCl   (4) HNO_3 and HCl   30 ____

Reference table K and L

PART B–1

Answer all questions in this part.

Directions (31–50): For *each* statement or question, write in the answer space the *number* of the word or expression that, of those given, best completes the statement or answers the question. Some questions may require the use of the *2011 Edition Reference Tables for Physical Setting/Chemistry.*

31 Which electron configuration represents the electrons of an atom of neon in an excited state?

(1) 2-7   (3) 2-7-1
(2) 2-8   (4) 2-8-1   31 _____

electrons move up a level when excited

32 Some information about the two naturally occurring isotopes of gallium is given in the table below.

**Natural Abundance of
Two Gallium Isotopes**

Isotope	Natural Abundance (%)	Atomic Mass (u)
Ga-69	60.11	68.926
Ga-71	39.89	70.925

Which numerical setup can be used to calculate the atomic mass of gallium?

(1) $(0.6011)(68.926 \text{ u}) + (0.3989)(70.925 \text{ u})$
(2) $(60.11)(68.926 \text{ u}) + (39.89)(70.925 \text{ u})$
(3) $(0.6011)(70.925 \text{ u}) + (0.3989)(68.926 \text{ u})$
(4) $(60.11)(70.925 \text{ u}) + (39.89)(68.926 \text{ u})$ 32 _____

Percentage × atomic mass + percentage × atomic mass

33 A student measures the mass and volume of a sample of copper at room temperature and 101.3 kPa. The mass is 48.9 grams and the volume is 5.00 cubic centimeters. The student calculates the density of the sample. What is the percent error of the student's calculated density?

(1) 7.4%   (3) 9.2%
(2) 8.4%   (4) 10.2%   33 _____

Reference table T

34 What is the chemical formula for sodium sulfate?

(1) Na_2SO_4 (3) $NaSO_4$

(2) Na_2SO_3 (4) $NaSO_3$ 34 _____

Look this up on table E

35 Given the balanced equation representing a reaction:

$$2Na(s) + Cl_2(g) \rightarrow 2NaCl(s) + energy$$

If 46 grams of Na and 71 grams of Cl_2 react completely, what is the total mass of NaCl produced?

(1) 58.5 g   (3) 163 g

(2) 117 g   (4) 234 g   35 _____

36 Given the balanced equation representing a reaction:

$$2NO + O_2 \rightarrow 2NO_2 + energy$$

The mole ratio of NO to NO_2 is

(1) 1 to 1   (3) 3 to 2

(2) 2 to 1   (4) 5 to 2   36 _____

Same amount of NO a NO2

37 The particle diagram below represents a solid sample of silver.

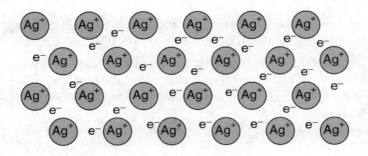

Which type of bonding is present when valence electrons move within the sample?

(1) metallic bonding   (3) covalent bonding

(2) hydrogen bonding   (4) ionic bonding   37 _____

Metals combine with metallic bonding

38 Given the formula representing a molecule:

$$H-\overset{\displaystyle H}{\underset{\displaystyle H}{C}}-\overset{\displaystyle H}{\underset{\displaystyle H}{C}}-\overset{\displaystyle H}{\underset{\displaystyle H}{C}}-H$$

Which statement explains why the molecule is nonpolar?

(1) Electrons are shared between the carbon atoms and the hydrogen atoms.
(2) Electrons are transferred from the carbon atoms to the hydrogen atoms.
(3) The distribution of charge in the molecule is symmetrical.
(4) The distribution of charge in the molecule is asymmetrical.   38 _____

Polarity = asymetry

39 A solid sample of a compound and a liquid sample of the same compound are each tested for electrical conductivity. Which test conclusion indicates that the compound is ionic?

(1) Both the solid and the liquid are good conductors.
(2) Both the solid and the liquid are poor conductors.
(3) The solid is a good conductor, and the liquid is a poor conductor.
(4) The solid is a poor conductor, and the liquid is a good conductor.   39 _____

40 Which statement explains why 10.0 mL of a 0.50 M H_2SO_4(aq) solution exactly neutralizes 5.0 mL of a 2.0 M NaOH(aq) solution?

(1) The moles of H^+(aq) equal the moles of OH^-(aq).
(2) The moles of H_2SO_4(aq) equal the moles of NaOH(aq).
(3) The moles of H_2SO_4(aq) are greater than the moles of NaOH(aq).
(4) The moles of H^+(aq) are greater than the moles of OH^-(aq).   40 _____

41 Which particle diagram represents *one* substance in the gas phase?

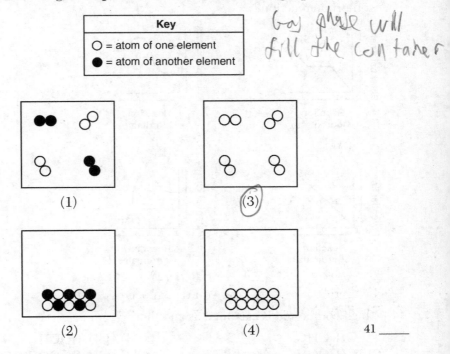

Gas phase will
fill the container

Key
O = atom of one element
● = atom of another element

(1)   (3)

(2)   (4)

41 _____

42 Given the equation representing a chemical reaction at equilibrium in a sealed, rigid container:

$$H_2(g) + I_2(g) + energy \rightleftharpoons 2HI(g)$$

When the concentration of $H_2(g)$ is increased by adding more hydrogen gas to the container at constant temperature, the equilibrium shifts

(1) to the right, and the concentration of HI(g) decreases
(2) to the right, and the concentration of HI(g) increases
(3) to the left, and the concentration of HI(g) decreases
(4) to the left, and the concentration of HI(g) increases

42 _____

Concentration shifts to be balanced

43 Which diagram represents the potential energy changes during an exothermic reaction?

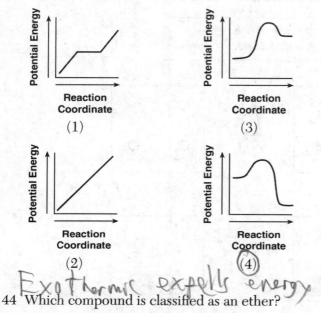

(1)

(3)

(2)

(4)

43 _____

Exothermic expells energy

44 Which compound is classified as an ether?

(1) CH_3CHO (3) CH_3COCH_3

(2) CH_3OCH_3 (4) CH_3COOCH_3 44 _____

Reference table R

45 Given the equation representing a reversible reaction:

$$HCO_3^-(aq) + H_2O(\ell) \rightleftharpoons H_2CO_3(aq) + OH^-(aq)$$

Which formula represents the H^+ acceptor in the forward reaction?

(1) $HCO_3^-(aq)$ (3) $H_2CO_3(aq)$

(2) $H_2O(\ell)$ (4) $OH^-(aq)$ 45 _____

$HCO_3^-(aq)$ become $H_2CO_3(aq)$

46 What is the mass of an original 5.60-gram sample of iron-53 that remains unchanged after 25.53 minutes?

(1) 0.35 g   (3) 1.40 g

(2) 0.70 g   (4) 2.80 g   46 _____

Divide by 2 for every half life

47 Given the equation representing a nuclear reaction:

$$_1^1H + X \rightarrow \, _3^6Li + \, _2^4He$$

The particle represented by X is

(1) $_4^9Li$

(3) $_5^{10}Be$

(2) $_4^9Be$ missing # of protons and mass   47 ____

48 Fission and fusion reactions both release energy. However, only fusion reactions

(1) require elements with large atomic numbers
(2) create radioactive products
(3) use radioactive reactants
(4) combine light nuclei   48 ____

Fusion fuses things

49 The chart below shows the crystal shapes and melting points of two forms of solid phosphorus.

Two Forms of Phosphorus

Form of Phosphorus	Crystal Shape	Melting Point (°C)
white	cubic	44
black	orthorhombic	610

Which phrase describes the two forms of phosphorus?

(1) same crystal structure and same properties
(2) same crystal structure and different properties
(3) different crystal structures and different properties
(4) different crystal structures and same properties   49 ____

50 Which graph shows the relationship between pressure and Kelvin temperature for an ideal gas at constant volume?

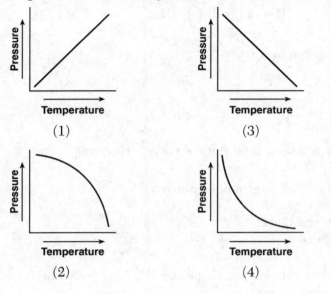

50 _____

PART B–2

Answer all questions in this part.

Directions (51–65): Record your answers on the answer sheet provided. Some questions may require the use of the *2011 Edition Reference Tables for Physical Setting/Chemistry.*

Base your answers to questions 51 through 53 on the information below and on your knowledge of chemistry.

The elements in Group 17 are called halogens. The word "halogen" is derived from Greek and means "salt former."

51　State the trend in electronegativity for the halogens as these elements are considered in order of increasing atomic number.　[1]

52　Identify the type of chemical bond that forms when potassium reacts with bromine to form a salt.　[1]

53　Based on Table *F*, identify *one* ion that reacts with iodide ions in an aqueous solution to form an insoluble compound.　[1]

Base your answers to questions 54 through 57 on the information below and on your knowledge of chemistry.

The diagrams below represent four different atomic nuclei.

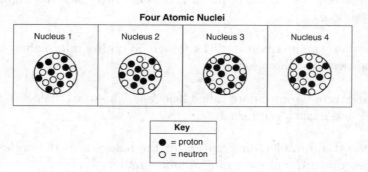

Four Atomic Nuclei

54 Identify the element that has atomic nuclei represented by nucleus 1.   [1]

55 Determine the mass number of the nuclide represented by nucleus 2.   [1]

56 Explain why nucleus 2 and nucleus 4 represent the nuclei of two different isotopes of the same element.   [1]

57 Identify the nucleus above that is found in an atom that has a stable valence electron configuration.   [1]

Base your answers to questions 58 through 60 on the information below and on your knowledge of chemistry.

The equation below represents a chemical reaction at 1 atm and 298 K.

$$2H_2(g) + O_2(g) \rightarrow 2H_2O(g)$$

58 State the change in energy that occurs in order to break the bonds in the hydrogen molecules. [1]

59 In the space *on your answer sheet*, draw a Lewis electron-dot diagram for a water molecule. [1]

60 Compare the strength of attraction for electrons by a hydrogen atom to the strength of attraction for electrons by an oxygen atom within a water molecule. [1]

Base your answers to questions 61 through 63 on the information below and on your knowledge of chemistry.

- A test tube contains a sample of solid stearic acid, an organic acid.
- Both the sample and the test tube have a temperature of 22.0°C.
- The stearic acid melts after the test tube is placed in a beaker with 320. grams of water at 98.0°C.
- The temperature of the liquid stearic acid and water in the beaker reaches 74.0°C.

61 Identify the element in stearic acid that makes it an organic compound.   [1]

62 State the direction of heat transfer between the test tube and the water when the test tube was placed in the water.   [1]

63 Show a numerical setup for calculating the amount of thermal energy change for the water in the beaker.   [1]

Base your answers to questions 64 and 65 on the information below and on your knowledge of chemistry.

A nuclear reaction is represented by the equation below.

$$_1^3\text{H} \rightarrow {}_2^3\text{He} + {}_{-1}^{0}\text{e}$$

64　Identify the decay mode of hydrogen-3.　[1]

65　Explain why the equation represents a transmutation.　[1]

PART C

Answer all questions in this part.

Directions (66–85): Record your answers on the answer sheet provided. Some questions may require the use of the *2011 Edition Reference Tables for Physical Setting/Chemistry*.

Base your answers to questions 66 through 68 on the information below and on your knowledge of chemistry.

A technician recorded data for two properties of Period 3 elements. The data are shown in the table below.

Two Properties of Period 3 Elements

Element	Na	Mg	Al	Si	P	S	Cl	Ar
Ionic Radius (pm)	95	66	51	41	212	184	181	—
Reaction with Cold Water	reacts vigorously	reacts very slowly	no observable reaction	no observable reaction	no observable reaction	no observable reaction	reacts slowly	no observable reaction

66 Identify the element in this table that is classified as a metalloid.   [1]

67 State the phase of chlorine at 281 K and 101.3 kPa.   [1]

68 State evidence from the technician's data which indicates that sodium is more active than aluminum.   [1]

Base your answers to questions 69 through 71 on the information below and on your knowledge of chemistry.

Ammonia, $NH_3(g)$, can be used as a substitute for fossil fuels in some internal combustion engines. The reaction between ammonia and oxygen in an engine is represented by the unbalanced equation below.

$$NH_3(g) + O_2(g) \rightarrow N_2(g) + H_2O(g) + \text{energy}$$

69 Balance the equation *on the answer sheet* for the reaction of ammonia and oxygen, using the smallest whole-number coefficients.   [1]

70 Show a numerical setup for calculating the mass, in grams, of a 4.2-mole sample of O_2. Use 32 g/mol as the gram-formula mass of O_2.   [1]

71 Determine the new pressure of a 6.40-L sample of oxygen gas at 300. K and 100. kPa after the gas is compressed to 2.40 L at 900. K.   [1]

Base your answers to questions 72 through 76 on the information below and on your knowledge of chemistry.

Fruit growers in Florida protect oranges when the temperature is near freezing by spraying water on them. It is the freezing of the water that protects the oranges from frost damage. When $H_2O(\ell)$ at 0°C changes to $H_2O(s)$ at 0°C, heat energy is released. This energy helps to prevent the temperature inside the orange from dropping below freezing, which could damage the fruit. After harvesting, oranges can be exposed to ethene gas, C_2H_4, to improve their color.

72 Write the empirical formula for ethene.   [1]

73 Explain, in terms of bonding, why the hydrocarbon ethene is classified as unsaturated.   [1]

74 Determine the gram-formula mass of ethene.   [1]

75 Explain, in terms of particle arrangement, why the entropy of the water *decreases* when the water freezes.   [1]

76 Determine the quantity of heat released when 2.00 grams of $H_2O(\ell)$ freezes at 0°C.   [1]

Base your answers to questions 77 through 80 on the information below and on your knowledge of chemistry.

A student constructs an electrochemical cell during a laboratory investigation. When the switch is closed, electrons flow through the external circuit. The diagram and ionic equation below represent this cell and the reaction that occurs.

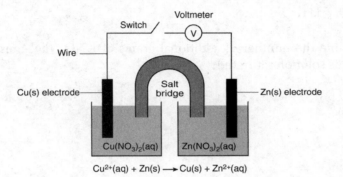

$$Cu^{2+}(aq) + Zn(s) \longrightarrow Cu(s) + Zn^{2+}(aq)$$

77 State the form of energy that is converted to electrical energy in the operating cell. [1]

78 State, in terms of the Cu(s) electrode and the Zn(s) electrode, the direction of electron flow in the external circuit when the cell operates. [1]

79 Write a balanced equation for the half-reaction that occurs in the Cu half-cell when the cell operates. [1]

80 State what happens to the mass of the Cu electrode and the mass of the Zn electrode in the operating cell. [1]

Base your answers to questions 81 and 82 on the information below and on your knowledge of chemistry.

A solution is made by dissolving 70.0 grams of $KNO_3(s)$ in 100. grams of water at 50.°C and standard pressure.

81 Show a numerical setup for calculating the percent by mass of KNO_3 in the solution.   [1]

82 Determine the number of additional grams of KNO_3 that must dissolve to make this solution saturated.   [1]

Base your answers to questions 83 through 85 on the information below and on your knowledge of chemistry.

Vinegar is a commercial form of acetic acid, $HC_2H_3O_2(aq)$. One sample of vinegar has a pH value of 2.4.

83 Explain, in terms of particles, why $HC_2H_3O_2(aq)$ can conduct an electric current.   [1]

84 State the color of bromthymol blue indicator in a sample of the commercial vinegar.   [1]

85 State the pH value of a sample that has ten times *fewer* hydronium ions than an equal volume of a vinegar sample with a pH value of 2.4.   [1]

Answer Sheet
June 2017
Chemistry—Physical Setting

PART B–2

51 As the atomic number of the Halogens increases, the electro negetivity decreases

52 Ionic bond

53 Lead (II) ion

54 Oxygen

55 18

56

57 Nucleus 3

58 Energy must be absorbed to break a covalent bond

59

60 Oxygen has a much stronger pull than the hydrogens because its electronegativity is higher

61 Carbon

62

63

$320 \cdot 4.18 \cdot - 24$

64 Beta

65 This is a transmutation because the hydrogen turns into helium

PART C

66 _Si is a metaloid_

67 _Gas_

68 _____

_____

_____

69 $\underline{2}$ $NH_3(g) + O_2(g) \rightarrow$ ___ $N_2(g) + \underline{3}$ $H_2O(g)$ + energy

70

$$4.2 = \frac{x}{32}$$

71 $\underline{800}$ **kPa** $\frac{(100)(6.4)}{300} = \frac{x(2.4)}{900}$

72 _CH2_

73 _Hydro carbon ethane has a double bond and_
could recive another hydrogen

74 _28_ g/mol

75 When water freezes, its entropy
decreases because its particles move
more slowly

76 _668_ J

77 Chemical

78 Electrons flow from the zinc
electrode to the copper electrode

79 $Cu^{2+} + 2e^- \rightarrow Cu^0$

80 Cu electrode: _increases_

Zn electrode: _decreases_

81

$$\frac{70}{120} \cdot 100$$

82 __15_____ g

83 Vinegar is an electrolyte, which means that in its aqueous form, it contains free flowing electrons

84 __yellow_____

85 __3.4_____

Answers
June 2017
Chemistry—Physical Setting

Answer Key

PART A

1. 3	**7.** 2	**13.** 1	**19.** 3	**25.** 2
2. 2	**8.** 1	**14.** 4	**20.** 4	**26.** 4
3. 1	**9.** 4	**15.** 1	**21.** 1	**27.** 3
4. 2	**10.** 4	**16.** 2	**22.** 3	**28.** 1
5. 4	**11.** 2	**17.** 4	**23.** 3	**29.** 2
6. 2	**12.** 1	**18.** 2	**24.** 1	**30.** 3

PART B–1

31. 3	**35.** 2	**39.** 4	**43.** 4	**47.** 2
32. 1	**36.** 1	**40.** 1	**44.** 2	**48.** 4
33. 3	**37.** 1	**41.** 3	**45.** 1	**49.** 3
34. 1	**38.** 3	**42.** 2	**46.** 2	**50.** 1

PART B–2 and **PART C**. *See* **Answers Explained**.

Answers Explained

PART A

1. **3** The charges of all subatomic particles should be committed to memory, but *just to be sure* you can consult Reference Table *O*, which shows the symbols, mass numbers, and charges of subatomic particles.

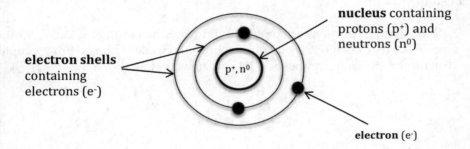

2. **2** In the wave-mechanical, or quantum, model of the atom, an orbital is defined as a three-dimensional region of space in which there is a high probability of locating an electron that has a certain energy.

3. **1** Atoms are neutral in charge, since the number of positively charged protons is equal to the number of negatively charged electrons. Reference Table *S* confirms that the element scandium has an atomic number of 21. An atomic number is defined as the number of protons in the nucleus of an atom. Since scandium atoms have 21 protons, they must also have 21 electrons.

4. **2** Atoms are built from three subatomic particles: protons, neutrons, and electrons. It stands to reason that all subatomic particles are lighter than the atoms they build. According to Reference Table *O*, a proton has a mass of approximately 1 amu, while an electron (which is equivalent to a beta particle) has a mass of 0 amu.
   (NOTE: Electrons *do* have mass, but when rounded to one significant figure, the mass of an electron is recorded as 0 amu.)

5. **4** Electrons that are located farther from the nucleus have more energy than those that are located closer to the nucleus. Energy is released when electrons move from higher energy shells to lower energy shells within the atom.

(NOTE: The first electron shell, or the first principal energy level, is the shell located closest to the nucleus.)

WRONG CHOICES EXPLAINED:
(1), (2), and (3) all describe electron transitions from a lower energy shell to a higher energy shell. These transitions would require the electron to absorb energy.

6. **2** The atomic number is defined as the number of protons in the nucleus of an atom.

7. **2** Structural formulas are drawn to illustrate the covalent bonding within a molecule. Lines drawn between atoms represent covalent bonds. For example, the structural formula for propene, C_3H_6, is shown below.

WRONG CHOICES EXPLAINED:
(1) Ionic compounds do not contain covalent bonds. Therefore, lines are never drawn between atoms in ionic formulas.
(3) Empirical formulas give the smallest whole-number ratio of atoms in a compound. For example, the empirical formula of propene, C_3H_6, is CH_2.
(4) Molecular formulas are the most common way of representing compounds. These formulas show how many atoms of each type of element are in a compound. In the case of propene, the molecular formula is C_3H_6.

8. **1** Chemical reactions occur when the chemical bonds in compounds are broken and then the atoms form new bonds, creating new compounds or elements. The products of a chemical reaction will have different physical properties than the reactants. Density, vapor pressure, and melting point are physical properties of materials that will change as new materials form. However, all chemical reactions involve the conservation of mass, energy, and charge.

9. **4** Double replacement reactions follow the general pattern AB + CD → AD + CB. AB and CD represent ionic compounds. During the course of the reaction, the positively charged cations (A and C) switch places and form new compounds with the opposite negatively charged anion.

10. **4** Ionic compounds differ widely in solubility, as seen on Reference Table G.

WRONG CHOICES EXPLAINED:

(1) The samples have the same mass, 5.0 grams, as stated in the question.

(2) Since density is the ratio of mass to volume $\left(d = \dfrac{m}{V}\right)$, if the density is the same and the mass is also the same between the two solids, then the solids must also occupy the same volume.

(3) Two very different solids can be at the same temperature, which indicates the amount of thermal energy in the material. The temperature of a sample does not provide information about the identity of the sample.

11. **2** The Lewis dot diagram for an I_2 molecule is shown below.

Since iodine atoms have seven valence electrons, each atom shares its unpaired electron with the other, giving each atom a stable octet of valence electrons. Two electrons are shared between iodine atoms, forming one covalent bond.

12. **1** Nonpolar covalent bonds are formed when electrons are shared between atoms that have the *same electronegativity*. In the Cl_2 molecule, both atoms have an equal attraction for the electrons in the bond, so partial negative and positive charges do not form.

13. **1** The ground-state electron configuration of a potassium atom is 2-8-8-1. A potassium atom must lose its valence electron to form the potassium ion, K^{+1}. Since it does not have any electrons in the fourth shell, the potassium ion is smaller than the potassium atom. Choices (3) and (4) cannot be correct because potassium atoms and potassium ions both contain 19 protons.

14. **4** Thermal energy is created by the motion of atoms and molecules within an object. The faster the particles move, the more thermal energy the object has and the hotter the object will be.

WRONG CHOICES EXPLAINED:

(1) Chemical energy refers to energy stored in chemical bonds.

(2) Electrical energy is the energy caused by a moving current of electrons.

(3) Nuclear energy is the energy released during nuclear reactions such as fission or fusion.

15. **1** Temperature is directly related to the average kinetic energy of particles in a sample of matter. If the temperature increases, then the average kinetic energy increases. Likewise, if the temperature decreases, then the average kinetic energy decreases. Kinetic energy is constant when the temperature does not change.

16. **2** The joule is a unit of energy in the metric system of measurement. Reference Table *D* provides a handy description of most of the units needed in the study of chemistry.

17. **4** Real gases behave more like ideal gases when temperature is high and pressure is low. These conditions minimize any possible attractions between gas particles, allowing the gas to adhere more closely to the principles of the Kinetic Molecular Theory of Gases.

WRONG CHOICES EXPLAINED:

(1) According to Reference Table *S*, helium gas condenses to a liquid at 4 K, which is far below standard temperature!

(2) Gases become less soluble at higher temperatures.

(3) Helium is a noble gas. Helium atoms never bond to other atoms.

18. **2** When gas particles are heated, the temperature of the sample increases and the particles move more quickly. This is consistent with the Kinetic Molecular Theory of Gases, which states that temperature is directly related to the average kinetic energy of the gas particles. Remember, kinetic energy is the energy of motion. Imagine that the particles in the container shown below increase in speed. Clearly the particles will strike each other more often, as well as the walls of the container, and since they are moving faster, they will collide with more force.

GAS

19. **3** This question is straightforward if you remember Avogadro's Law, which is shown below as a mathematical expression.

$$V \varpropto n$$

This law states that the volume of a gas is proportional to the number of particles in a gas sample if the temperature and pressure of the sample are held constant. Simply put, this law means that as the number of particles doubles, the volume doubles. However, if the number of particles is the same, as stated in this question, the volumes must also be the same for the two samples.

20. **4** Mixtures are defined as materials with variable composition containing two or more substances. This distinguishes mixtures from elements and compounds, which are pure substances with definite compositions.

WRONG CHOICES EXPLAINED:
(1), (2), and (3) are all true statements, but none of these statements explains why there are different percentages of hydrocarbons in different samples of crude oil.

21. **1** Most solids are crystalline solids, which mean that the particles are arranged in an orderly, geometric pattern. $CaCl_2$ is the only solid listed among the answer choices.

22. **3** Chemical equilibrium is established when the *rates* of the forward and reverse reactions are equal, NOT the amounts of products and reactants. Imagine cars traveling over a bridge at the same rate in both directions. The number of cars on each side of the bridge would remain constant, but the number of cars on each side of the bridge may not be equal. The condition of chemical equilibrium is illustrated below, showing that the concentration of products and the concentration of reactants become constant at equilibrium.

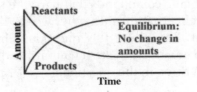

23. **3** The heat of reaction (ΔH) is defined as the potential energy of the products of a chemical reaction minus the potential energy of the reactants. In other words, the heat of reaction is the net change in energy during the reaction. This is a definition that must be memorized. The potential energy diagram below can help you visualize this definition.

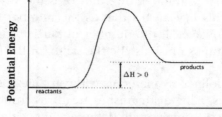

24. **1** Many chemical reactions proceed from reactants to products in a series of steps that involve particle collisions rather than as one single collision. A catalyst creates an alternate sequence of steps that leads to a lower activation energy for the complete reaction. Having a lower activation energy overall leads to a faster reaction since more particles will have sufficient kinetic energy to react.

25. **2** Reference Table *R* lists general structural formulas and examples for some very common organic functional groups. Examine this table and look for nitrogen-containing functional groups. You will see that this compound must be either an amine or an amide. Further examination shows that this compound must be classified as an amine since it has the functional group $-NH_2$. There are three carbons in this amine, so by using the prefixes listed on Reference Table *P*, you will see that this results in the name 1-propanamine. Incidentally, the compound in this question is the exact same as the example given on Reference Table *R*. **Use your Reference Tables!**

26. **4** Isomers are organic compounds that have the same numbers and types of atoms but have different structures. In other words, the atoms are attached in a different arrangement. To answer this question, simply choose the answer choice that has 4 carbon atoms, 10 hydrogen atoms, and 1 oxygen atom, just like the original compound written in the question.

27. **3** A reaction between any organic acid and an alcohol is known as an esterification reaction. See the example below.

$$
\begin{array}{ccccccc}
\text{O} & & & & \text{O} & & \\
\parallel & & & & \parallel & & \\
\text{RC—OH} & + & \text{R}'\text{OH} & \xrightarrow{\Delta} & \text{RC—OR}' & + & \text{H}_2\text{O} \\
\text{carboxylic acid} & & \text{alcohol} & & \text{ester} & & \text{water}
\end{array}
$$

28. **1** Redox reactions involve the transfer of electrons. In these reactions, as in all reactions, charge must be conserved. In practice, this means that the number of electrons lost by the substance that oxidizes must equal the number of electrons gained by the substance that reduces.

29. **2** The process of electrolysis utilizes an electrolytic cell that consumes electrical energy to force a nonspontaneous chemical change to occur. For example, highly reactive alkali metals, such as sodium, can be generated in their pure state using electrolysis to drive an electron back onto alkali metal ions.

WRONG CHOICES EXPLAINED:
  (1) Deposition is a phase change from gas to solid directly. Phase changes are always physical changes, not chemical changes.
  (3) Alpha decay is a form of radioactive decay, which is a nuclear reaction.
  (4) Chromatography is a separation technique for liquid or gaseous mixtures.

30. **3** Reference Tables K and L list common acids and bases. Reference Table K lists HNO_3 (nitric acid) and Reference Table L lists $NaOH$ (sodium hydroxide). Arrhenius acids release H^+ ions in aqueous solution, and Arrhenius bases release OH^- ions in aqueous solution.

WRONG CHOICES EXPLAINED:
  (1) Neither of these compounds is an acid. CH_3OH is an organic compound known as an alcohol.
  (2) Neither of these compounds is a base. CH_3OH is an organic compound known as an alcohol. HCl is an acid.
  (4) Both of these compounds are acids listed on Reference Table K.

PART B–1

31. **3** An atom of neon will always have 10 electrons, regardless of whether it is in the ground state or in the excited state. The Periodic Table in the Reference Tables shows that the ground-state electron configuration of a neon atom is 2-8. A neon atom would be in an excited state if its configuration were 2-7-1 because this configuration shows that *a second-shell electron has moved to the third shell.*

WRONG CHOICES EXPLAINED:
    (1), (4) These configurations do not include 10 electrons.
    (2) This is the ground-state, lowest energy electron configuration for neon atoms.

32. **1** The atomic mass of an element is found by taking the weighted average of all naturally occurring isotopes of that element. The natural abundance percentages of each isotope are first converted to decimal fractions, and these are then multiplied by the atomic mass of the isotope. Adding the contributions of each isotope together yields the correct atomic mass for that element.

33. **3** The density and percent error formulas can be found on Reference Table *T*. Using the information given in the question, first calculate the density of copper using the student's mass and volume measurements:

$$d = \frac{m}{V}$$

$$d = \frac{48.9 \text{ g}}{5.00 \text{ cm}^3}$$

$$d = 9.78 \text{ g/cm}^3$$

The next step is to calculate the student's percent error for the calculated value shown above by comparing it to the accepted value for the density of copper that is listed on Reference Table *S*.

$$\text{percent error} = \frac{\text{measured value} - \text{accepted value}}{\text{accepted value}} \times 100$$

Plug the calculated value of 9.78 g/cm^3 and the accepted value of 8.96 g/cm^3 into the formula:

$$\text{percent error} = \frac{9.78 \text{ g/cm}^3 - 8.96 \text{ g/cm}^3}{8.96 \text{ g/cm}^3} \times 100$$

$$\text{percent error} = 9.15\% \approx 9.2\%$$

34. **1** Sodium sulfate is an ionic compound made from the sodium ion, Na^{+1}, and the sulfate ion, SO_4^{-2}. Charges for these two ions can be found on the Periodic Table of the Elements and on Reference Table E, respectively. Since compounds have no overall charge, there must be two sodium ions present in the compound to cancel the -2 charge of the sulfate ion.

WRONG CHOICES EXPLAINED:
(2) This is the formula for sodium sulf*ite*. Look carefully at Reference Table E to notice the difference between the formulas of the sulfite and the sulfate ions.

(3), (4) These formulas are incorrect since the sum of ion charges is not neutral.

35. **2** This answer depends on simple arithmetic involving the Law of Conservation of Mass. Matter is neither created nor destroyed in chemical reactions. The total mass of the products must always equal the mass of the reactants. The mass of both reactants is given in the question. Adding both together will equal the mass of the NaCl produced, as shown below:

$$2Na(s) + Cl_2(g) \rightarrow 2NaCl(s) + \text{energy}$$
$$46 \text{ g} + 71 \text{ g} = 117 \text{ g}$$

36. **1** The coefficients in a balanced chemical equation represent moles or particles. Both NO and NO_2 have a coefficient of 2. A ratio of 2:2 expressed in lowest terms is a 1:1 relationship.

37. **1** Metallic bonds hold samples of pure metals together. Metal atoms lose valence electrons to a common "sea of electrons." This mobile pool of delocalized electrons serves as a sort of "glue" for the metal cations created when metal atoms lose their valence electrons.

38. **3** Nonpolar molecules do not have partial positive or negative charges. This is because the charge is distributed symmetrically throughout the molecule. The propane molecule has a symmetrical arrangement of atoms in which a central chain of three carbon atoms is symmetrically surrounded by hydrogen atoms. There is no location on this molecule with greater or lesser electronegativity than any other location, so partial charges do not occur.

39. **4** Ionic solids do not conduct electricity because the ions are locked in place within a crystalline lattice structure. Ions are charged particles and can assist with conduction only if the ions are free to move. Melting the ionic compound creates mobile ions, so the molten compound can conduct electricity.

40. **1** A neutralization reaction involves H^+ ions from an acid reacting with OH^- ions from a base, forming water. These ions must be present in exactly equal amounts for perfect neutralization to occur.

41. **3** A pure substance is represented by just one type of particle throughout the sample. For a pure gas sample, the particles spread out to fill the entire space, just as gases do. The correct answer represents a diatomic gas, like pure O_2 or N_2.

WRONG CHOICES EXPLAINED:
(1) This particle diagram shows a mixture of two different diatomic gases.
(2), (4) These particle diagrams show the regular arrangement of particles in the solid phase.

42. **2** Systems at equilibrium prefer to remain at equilibrium. If additional $H_2(g)$ is added to this system, the reaction "shifts to the right" to use up the extra H_2 that was added. This results in the production of extra $HI(g)$, so when a new equilibrium is established, there will be more $HI(g)$ present than there was initially.

43. **4** Energy is released, or lost, during exothermic chemical reactions, causing these reactions to have a $-\Delta H$ (heat of reaction). The products of these reactions contain *less* stored energy in their bonds than the energy stored in the bonds of the reactants. In order to form the lower energy products, energy must be released.

44. **2** Refer to Reference Table *R* for all questions that involve organic functional groups. Ethers contain an oxygen atom singly bonded to two carbon atoms in the middle of a hydrocarbon chain.

WRONG CHOICES EXPLAINED:

(1) This compound is an aldehyde, with the oxygen doubly bonded to a carbon at the end of the molecule.

(3) This compound is a ketone. The oxygen atom creates a double bond with a carbon atom in the middle of the molecule.

(4) This compound is an ester, with two oxygens in the middle of the molecule.

45. **1** The Brønsted–Lowry theory of acids and bases defines an acid as a proton (H^+) donor and a base as a proton (H^+) acceptor. $HCO_3^-(aq)$ must accept an H^+ in the forward reaction in order to form the product H_2CO_3.

46. **2** From Reference Table N, the half-life of Fe-53 is 8.51 minutes. In 25.53 minutes, Fe-53 goes through 3 half-lives since 25.53 minutes divided by 8.51 minutes equals 3 half-lives. This question is really asking how much Fe-53 will remain in the sample after 3 half-lives have passed. The mass of a radioisotope reduces by half for each half-life that passes. This problem is solved by taking the original mass and dividing it in half 3 successive times, as shown below:

$$1\text{st half-life:} \quad \frac{5.60 \text{ grams}}{2} = 2.80 \text{ grams}$$

$$2\text{nd half-life:} \quad \frac{2.80 \text{ grams}}{2} = 1.4 \text{ grams}$$

$$3\text{rd half-life:} \quad \frac{1.4 \text{ grams}}{2} = 0.70 \text{ grams}$$

47. **2** To balance a nuclear equation, the sum of the atomic numbers (bottom) and the mass numbers (top) must be equal on both sides of the arrow. A mass number of 9 and an atomic number of 4 correctly balances this nuclear equation. Since the atomic number is 4, the element is beryllium, Be.

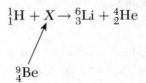

$$^1_1H + X \rightarrow {}^6_3Li + {}^4_2He$$

$$^9_4Be$$

48. **4** Fusion is a nuclear reaction in which two very lightweight atomic nuclei, usually hydrogen, are combined into a slightly heavier nucleus, usually helium.

WRONG CHOICES EXPLAINED:
   (1) Fusion takes place between elements with small atomic numbers.
   (2) Both fission and fusion produce radioactive products.
   (3) The reactants for nuclear fusion do not need to be radioactive.

49. **3** The chart supplied with this question contains the answer! The two forms of phosphorus have different crystal structures—one is cubic and the other is orthorhombic. The two forms also have different properties, such as the very different melting points shown.

50. **1** Increasing the Kelvin temperature increases the speed of the gas particles in the sample. Faster moving particles will collide more often and with more force against the inside of a container, causing the internal pressure to increase. This is a direct relationship.

PART B–2

[All questions in Part B–2 are worth 1 point.]

51. Electronegativity values can be found on Reference Table S. Simply look up the electronegativity values of the halogens as listed in Group 17 from the top of the group (low atomic number) to the bottom of the group (high atomic number):

Atomic Number	Element	Electronegativity
9	F	4.0
17	Cl	3.2
35	Br	3.0
53	I	2.7
85	At	2.2

Electronegativity decreases as the halogens are considered in order of increasing atomic number.

52. Ionic bonds form when metals bond with nonmetals. Potassium is a metal, and bromine is a nonmetal, so the bond between them would be an **ionic bond**.

53. Three exceptions are listed for the solubility of halide ions (Cl^-, Br^-, and I^-) in Reference Table F: the **silver ion (Ag^+), the lead (II) ion (Pb^{2+}), and the mercury (I) ion ($Hg_2{}^{2+}$)**. Any one of these is an acceptable answer to this question.

54. The number of protons in an atom (known as the atomic number) identifies the element. There are 9 protons drawn in Nucleus 1, so this must be the element **fluorine (F)**, which has an atomic number of 9.

55. The mass number is defined as the sum of the protons and neutrons in an atom. Each proton and neutron has an approximate mass of 1 amu, so the sum of these particles will be equal to the mass expressed in amu (atomic mass units). There are 8 protons and 10 neutrons drawn in Nucleus 2, so the mass number of this atom would be $8 + 10 = $ **18 (amu)**.

  (NOTE: Units were not required to receive full credit on this question.)

56. **Nucleus 2 and Nucleus 4 contain the same number of protons (8) but different numbers of neutrons (10 and 11, respectively).** Atoms of the same element that contain the same number of protons but different numbers of neutrons are called isotopes of that element.

57. The noble gases in Group 18 have a stable valence electron configuration, which is the reason why the noble gases do not tend to form chemical bonds. **Nucleus 3** contains 10 protons, and therefore represents an atom of neon, Ne, which is a noble gas.

58. **Energy must be absorbed** to break the covalent bond in the H_2 molecule, just as energy must be used to break apart two magnets that are stuck together.

59. Lewis electron-dot diagrams are drawn to show how the valence electrons of all atoms are distributed within the compound. These diagrams are used to illustrate the number and types of bonds in a compound. The Lewis electron-dot diagram for water is shown below. Note that in the finished diagrams, the oxygen atom achieves a stable octet of valence electrons by sharing each hydrogen's single valence electron.

$$\text{H}-\overset{\displaystyle ..}{\underset{\displaystyle ..}{\text{O}}}-\text{H} \quad or \quad \text{H}\,\overset{\displaystyle ..}{\underset{\displaystyle ..}{\text{:O:}}}\,\text{H}$$

60. To answer this question correctly, you need to recall two things: (1) electronegativity is defined as an atom's attraction for an electron in a covalent bond, and (2) values for electronegativity can be found on Reference Table S. Looking at Reference Table S, we find that **oxygen has a much higher electronegativity (3.4) than hydrogen (2.2).**

61. The element **carbon**, **C**, is the central atom in organic compounds. Organic chemistry is the study of carbon-containing compounds.

62. Heat always flows from substances at a higher temperature to substances at a lower temperature. In this case, **heat will flow from the water** (at a temperature of 98.0 °C) **to the test tube** (at a temperature of 22.0 °C).

63. "Thermal energy change" is just a fancy way of saying "amount of heat lost or gained." Heat formulas can be found on Reference Table T. Since the temperature of the water in the beaker decreases from 98.0 °C to 74.0 °C, the correct formula is:

$$q = mC\Delta T$$

In this formula, q = the amount of heat in joules (J) lost or gained, m = the mass of the water in grams, C = specific heat capacity of liquid water (as found on Reference Table B), and $\Delta T = T_{final} - T_{initial}$, which equals $74.0 - 98.0 = -24.0°C = -24.0$ K.

Plugging in values from the information given leads to the following expression:

$$q = (320. \text{ g}) (4.18 \text{ J/g} \cdot \text{K}) (-24.0 \text{ K})$$

(NOTE: Credit was given for both a + ΔT value or a – ΔT value.)

64. Nuclear particle symbols are found on Reference Table O. The nuclear equation in this question represents **beta decay** since a **beta particle**, $\beta^- \left({}^{\,\,0}_{-1}e \right)$ is emitted.

(NOTE: Acceptable answers included the term "beta decay" or either symbol for a beta particle.)

65. "Transmutation" means the conversion of one element into another by a nuclear process. The equation shown in this question represents transmutation because a **H-3 nucleus is converted to a He-3 nucleus**.

PART C

[All questions in Part C are worth 1 point.]

66. The six metalloids must be memorized. They are B, Si, As, Te, Sb, and Ge. These metalloids are easily remembered visually—just look at the positions of these elements on the Periodic Table and memorize the physical pattern that they make. The only metalloid listed in the table for this question is **Si**, or **silicon**.

67. Boiling point information for most of the elements on the Periodic Table can be found on Reference Table *S*. The asterisk for the "Boiling Point" column heading signifies that the boiling point data listed is for standard pressure, which is equal to 101.3 kPa. We can therefore use the information from Reference Table *S* to figure out this question. Since the boiling point for chlorine listed on Reference Table *S* is 239 K, chlorine would be well *above* its boiling point at 281 K and would therefore be a **gas**.

68. Answering this question only requires reading and interpreting the information printed in the table. **Sodium reacts vigorously with cold water whereas aluminum has no observable reaction.** If you struggle to remember which element corresponds to each symbol, remember that most elements and their symbols are listed on Reference Table *S*.

69. The basic idea when balancing chemical equations can be thought of as "Atoms In = Atoms Out." First, notice that there are two N atoms on the right side of the equation and only one N atom on the left side of the equation. A coefficient of 2 can be written in front of the NH_3 to balance the N atoms. With a coefficient of 2 in front of NH_3, there are 6 H atoms on the reactant side, so a coefficient of 3 is written in front of H_2O on the product side to balance the H atoms. The partially balanced equation is now shown below:

$$\underline{\quad 2 \quad} NH_3(g) + \underline{\quad\quad} O_2(g) \rightarrow \underline{\quad\quad} N_2(g) + \underline{\quad 3 \quad} H_2O(g) + \text{ energy}$$

We now run into a classic difficulty when balancing equations that contain diatomic elements, like O_2. Three oxygen atoms are now on the right of this equation, and yet they are only available as pairs on the left side of the equation. In a situation like this, *double all of the coefficients* (except the coefficient for the O_2, which isn't known yet). Doubling an odd number results in an even number. The finished balanced equation becomes:

$$\underline{\quad 4 \quad} \textbf{NH}_3(\textbf{g}) + \underline{\quad 3 \quad} \textbf{O}_2(\textbf{g}) \rightarrow \underline{\quad 2 \quad} \textbf{N}_2(\textbf{g}) + \underline{\quad 6 \quad} \textbf{H}_2\textbf{O}(\textbf{g}) + \textbf{energy}$$

70. The formula that relates moles to mass is found on Reference Table *T*.

$$\text{number of moles} = \frac{\text{given mass}}{\text{gram-formula mass}}$$

Plug the information given in this question into this formula to write the correct numerical setup:

$$4.2 \text{ mol} = \frac{x \text{ grams}}{32 \text{ g/mol}}$$

71. Use the combined gas law formula found on Reference Table T:

$$\frac{P_1 V_1}{T_1} = \frac{P_2 V_2}{T_2}$$

Substituting information from the problem:

$$\frac{(100. \text{ kPa})(6.40 \text{ L})}{300. \text{ K}} = \frac{(P_2)(2.40 \text{ L})}{900. \text{ K}}$$

Rearranging:

$$P_2 = \frac{(100. \text{ kPa})(6.40 \text{ L})(900. \text{ K})}{(300. \text{ K})(2.40 \text{ L})} = \textbf{800. kPa}$$

72. An empirical formula shows the elements of a compound in their lowest whole number ratio. The molecular formula for ethene, C_2H_4, can be reduced to lowest terms by dividing each subscript by 2. The empirical formula of ethene is $\textbf{CH}_2$.

73. Unsaturated hydrocarbons contain at least one double or triple bond between carbon atoms. The term "unsaturated" implies that other atoms could be added to the compound by breaking the multiple bond between the carbon atoms. The structural formula for ethene is shown below. **Ethene is classified as an unsaturated hydrocarbon because of the double bond between the carbon atoms.**

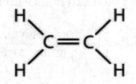

74. The gram-formula mass, also known as the molar mass, can be calculated by adding the molar masses of all of the elements in a compound. You can find the molar masses of each element on the Periodic Table provided in your Reference Tables.

Algebraically, the gram-formula mass for ethene can be expressed as:

$$[2 \times (\text{molar mass of carbon})] + [4 \times (\text{molar mass of hydrogen})]$$

Plug in the molar masses of each element and solve:

$$(2 \times 12.0) + (4 \times 1.0) = \textbf{28.0 g/mol}$$

75. Entropy is the measure of randomness or disorder in a system. When liquid water freezes into ice, independent, freely moving water molecules become locked into a rigid, geometric, crystalline structure. Entropy decreases during this process because **the molecules in liquid water are arranged more randomly than molecules in ice**.

(NOTE: This problem asks for the explanation to be written *in terms of particle arrangement*. To receive credit, your answer must mention the changing particle arrangement as liquid water freezes.)

76. The last page of the Reference Tables, Reference Table *T*, lists the formula to find the heat of fusion, which is the heat lost or gained by a substance as it freezes or melts, respectively.

$$q = mH_f$$

In this formula, q = heat, m = mass, and H_f = the heat of fusion.

The mass is given as 2.00 grams, and the heat of fusion for liquid water, found on Reference Table *B*, is 334 J/g. Plug these values into the heat of fusion formula and solve:

$$q = (2.00 \text{ g})(334 \text{ J/g}) = \textbf{668 J}$$

77. The electrochemical cell shown, known as a battery, converts **chemical energy** (the potential energy stored in chemical bonds) into electrical energy.

78. To decide the direction of electron flow in this system, you must determine which metal is more active—Zn or Cu. Reference Table *J* lists metals from the most active metals at the top to the least active metals at the bottom. A more active metal will oxidize (lose electrons) in the presence of a less active metal. Since zinc is higher than copper on Reference Table *J*, zinc will lose electrons to copper, causing electrons to flow **from the zinc electrode to the copper electrode** in this system.

79. Copper is lower than zinc on Reference Table *J*, so copper ions will *reduce* in the presence of zinc. Remember that metal *atoms* will never gain electrons to form negative ions—their electronegativity is too low. However, metal *ions* can regain electrons lost previously to re-form the original metal atoms. Recognizing that reduction happens in the Cu half-cell is critical for writing the correct reduction half-reaction for the cell, which is:

$$Cu^{2+} + 2e^- \rightarrow Cu^0$$

[NOTE: Credit was also awarded if the product was written as Cu(s) or even just Cu.]

80. **The mass of the Cu electrode increases, and the mass of the Zn electrode decreases.** This happens because soluble zinc ions form and dissolve as the zinc electrode loses electrons, and insoluble copper atoms form as copper ions in the copper half-cell gain electrons. The copper atoms formed are deposited right onto the copper electrode.

81. Use the percent composition formula found on Reference Table *T*:

$$\% \text{ composition by mass} = \frac{\text{mass of part}}{\text{mass of whole}} \times 100$$

When applying this formula to a solution, the "part" will always be the dissolved solute, while the "whole" will be the total mass of the solution (solute + solvent). Using the information given in this question, the total mass of the solution described is 70.0 g KNO_3 + 100. g H_2O = 170. g solution.

To write the numerical setup, simply plug the appropriate values into the percent composition formula:

$$\% \text{ composition by mass} = \frac{70.0 \text{ g } KNO_3}{170. \text{ g solution}} \times 100$$

82. Saturated solutions hold the maximum possible dissolved solute at a particular temperature. Reference Table G shows lines of saturation for several solutes. To solve this problem, first read the solubility in a saturated solution of KNO_3 at 50 °C, which is approximately 85. g KNO_3 in 100 g of water.

Therefore 85. g − 70. g = **15 additional grams of KNO$_3$** must be added to saturate this solution.

(NOTE: Credit was awarded for answers ranging from 12 g of KNO_3 to 16 g of KNO_3.)

83. Electrolytes are compounds that dissociate into ions in aqueous solution. The freely moving charged ions facilitate the flow of electricity through the resulting solution. Acids, bases, and salts (ionic compounds) are three classes of compounds that are electrolytes. Acetic acid solution, $HC_2H_3O_2(aq)$, can conduct an electric current because **it contains mobile, charged ions**.

84. Use Reference Table M to answer this question about an acid–base indicator. Bromthymol blue changes color from yellow to blue in the pH range of 6.0–7.6. Below a pH of 6.0, bromthymol blue is yellow in aqueous solution. Since vinegar has a pH of 2.4, vinegar would be **yellow** in the presence of bromthymol blue indicator.

85. pH is a function of the base 10 exponent of the hydronium ion concentration in an aqueous solution. As an example, a solution with a hydronium concentration of 0.001 M (which equals 10^{-3} M) would have a pH of −(−3), in other words a pH of 3. Likewise, a solution with a hydronium concentration of 0.0001 M (which equals 10^{-4} M) has a pH of 4. Each number on the pH scale differs in hydronium ion concentration by a factor of 10 from the next whole number higher and lower on the scale. In this problem, a sample that has 10 times *fewer* hydronium ions would differ by one whole number from the pH of the original vinegar. Having fewer hydronium ions means that the sample is less acidic than the original vinegar and would have a higher pH value. Putting this information together, the pH of the new sample would be 2.4 + 1.0 = **3.4**.

Mark (✔) the questions you answered correctly. Count the number of checks and follow the formulas given to determine your score on each topic.

Core Area	☐ Questions Answered Correctly

33: (1)

Section M—Math Skills
☐ Number of checks ÷ 1 × 100 = ____ %

16, 49, 68: (3)

Section R—Reading Skills
☐ Number of checks ÷ 3 × 100 = ____ %

1–5, 31, 32, 55, 56: (9)

Section I—Atomic Concepts
☐ Number of checks ÷ 9 × 100 = ____ %

6, 51, 54, 66: (4)

Section II—Periodic Table
☐ Number of checks ÷ 4 × 100 = ____ %

7–9, 34–36, 69, 70, 72, 74: (10)

Section III—Moles/Stoichiometry
☐ Number of checks ÷ 10 × 100 = ____ %

11–13, 37, 38, 52, 53, 57–60: (11)

Section IV—Chemical Bonding
☐ Number of checks ÷ 11 × 100 = ____ %

10, 14, 15, 17–21, 39, 41, 50, 62, 63, 67, 71, 76, 77, 81, 82: (19)

Section V—Physical Behavior of Matter
☐ Number of checks ÷ 19 × 100 = ____ %

22–24, 42, 43, 75: (6)

Section VI—Kinetics and Equilibrium
☐ Number of checks ÷ 6 × 100 = ____ %

25–27, 44, 61, 73: (6)

Section VII—Organic Chemistry
☐ Number of checks ÷ 6 × 100 = ____ %

Core Area	☐ Questions Answered Correctly

28, 29, 78–80: (5)

Section VIII—Oxidation–Reduction
☐ Number of checks ÷ 5 × 100 = ___ %

30, 40, 45, 83–85: (6)

Section IX—Acids, Bases, and Salts
☐ Number of checks ÷ 6 × 100 = ___ %

25–27, 44, 61, 73: (6)

Section X—Nuclear Chemistry
☐ Number of checks ÷ 5 × 100 = ___ %

Examination August 2017

Chemistry—Physical Setting

PART A

Answer all questions in this part.

Directions (1–30): For *each* statement or question, write in the answer space the *number* of the word or expression that, of those given, best completes the statement or answers the question. Some questions may require the use of the *2011 Edition Reference Tables for Physical Setting/Chemistry*.

1 Which phrase describes an Al atom?
  (1) a negatively charged nucleus, surrounded by negatively charged electrons
  (2) a negatively charged nucleus, surrounded by positively charged electrons
  (3) a positively charged nucleus, surrounded by negatively charged electrons
  (4) a positively charged nucleus, surrounded by positively charged electrons   1 _____

2 What is the number of electrons in an atom that has 20 protons and 17 neutrons?
  (1) 37   (3) 3
  (2) 20   (4) 17   2 _____

3 The mass of a proton is approximately equal to the mass of
  (1) an electron   (3) an alpha particle
  (2) a neutron   (4) a beta particle   3 _____

4 When a sample of $CO_2(s)$ becomes $CO_2(g)$, there is a change in

  (1) bond type

  (2) gram-formula mass

  (3) molecular polarity

  (4) particle arrangement   4 _____

5 Which properties are characteristic of Group 2 elements at STP?

  (1) good electrical conductivity and electronegativities less than 1.7

  (2) good electrical conductivity and electronegativities greater than 1.7

  (3) poor electrical conductivity and electronegativities less than 1.7

  (4) poor electrical conductivity and electronegativities greater than 1.7   5 _____

6 Compared to an atom of C-12, an atom of C-14 has a greater

  (1) number of electrons

  (2) number of protons

  (3) atomic number

  (4) mass number   6 _____

7 Elements that have atoms with stable valence electron configurations in the ground state are found in

  (1) Group 1

  (2) Group 8

  (3) Group 11

  (4) Group 18   7 _____

8 A magnesium atom that loses two electrons becomes a

  (1) positive ion with a smaller radius

  (2) negative ion with a smaller radius

  (3) positive ion with a larger radius

  (4) negative ion with a larger radius   8 _____

9 An atom of which element has the strongest attraction for the electrons in a bond?

  (1) aluminum

  (2) carbon

  (3) chlorine

  (4) lithium   9 _____

10 Which type of matter *cannot* be broken down into simpler substances by a chemical change?

(1) an element   (3) a mixture
(2) a solution   (4) a compound   10 _____

11 According to Table *F*, which substance is most soluble in water?

(1) $AgCl$ (3) Na_2CO_3
(2) $CaCO_3$ (4) $SrSO_4$ 11 _____

12 Given the equation representing a reaction:

$$H + H \rightarrow H_2$$

Which statement describes the energy change in this reaction?

(1) A bond is broken as energy is absorbed.
(2) A bond is broken as energy is released.
(3) A bond is formed as energy is absorbed.
(4) A bond is formed as energy is released.   12 _____

13 Which sample of matter is a mixture?

(1) air   (3) manganese
(2) ammonia   (4) water   13 _____

14 Paper chromatography can separate the components of a mixture of colored dyes because the components have differences in

(1) decay mode   (3) ionization energy
(2) thermal conductivity   (4) molecular polarity 14 _____

15 At standard pressure, the boiling point of an unsaturated $NaNO_3(aq)$ solution increases when

(1) the solution is diluted with water
(2) some of the $NaNO_3(aq)$ solution is removed
(3) the solution is stirred
(4) more $NaNO_3(s)$ is dissolved in the solution   15 _____

16 Which term identifies a form of energy?

(1) combustion
(3) thermal
(2) exothermic
(4) electrolytic   16 _____

17 According to kinetic molecular theory, which statement describes one characteristic of an ideal gas system?

(1) The distance between gas molecules is smaller than the diameter of one gas molecule.
(2) The attractive force between two gas molecules is strong.
(3) The energy of the system decreases as gas molecules collide.
(4) The straight-line motion of the gas molecules is constant and random.   17 _____

18 The temperature of a substance is a measure of the

(1) average kinetic energy of its particles
(2) average potential energy of its particles
(3) ionization energy of its particles
(4) activation energy of its particles   18 _____

19 A real gas behaves most like an ideal gas at

(1) low pressure and high temperature
(2) low pressure and low temperature
(3) high pressure and high temperature
(4) high pressure and low temperature   19 _____

20 A reaction is most likely to occur when the colliding particles have proper orientation and

(1) mass
(3) half-life
(2) volume
(4) energy   20 _____

21 At STP, a 12.0-liter sample of $CH_4(g)$ has the same total number of molecules as

(1) 6.0 L of $H_2(g)$ at STP
(2) 12.0 L of $CO_2(g)$ at STP
(3) 18.0 L of $HCl(g)$ at STP
(4) 24.0 L of $O_2(g)$ at STP   21 _____

22 At standard pressure, during which physical change does the potential energy decrease?

    (1) liquid to gas   (3) solid to gas

    (2) liquid to solid   (4) solid to liquid   22 _____

23 Which equation represents a chemical equilibrium?

    (1) $N_2(\ell) \rightleftharpoons N_2(g)$ (3) $CO_2(s) \rightleftharpoons CO_2(g)$

    (2) $2NO_2(g) \rightleftharpoons N_2O_4(g)$ (4) $NH_3(\ell) \rightleftharpoons NH_3(g)$ 23 _____

24 The amount of randomness of the atoms in a system is an indication of the

    (1) entropy of the system

    (2) polarity of the system

    (3) excited state of the atoms

    (4) ground state of the atoms   24 _____

25 When a sample of $Ca(s)$ loses 1 mole of electrons in a reaction with a sample of $O_2(g)$, the oxygen

    (1) loses 1 mole of electrons

    (2) loses 2 moles of electrons

    (3) gains 1 mole of electrons

    (4) gains 2 moles of electrons   25 _____

26 Which reaction occurs at the anode of an electro-chemical cell?

    (1) oxidation   (3) neutralization

    (2) reduction   (4) transmutation   26 _____

27 Which substance is an electrolyte?

    (1) CCl_4 (3) SiO_2

    (2) $C_6H_{12}O_6$ (4) H_2SO_4 27 _____

28 In which process does a heavy nucleus split into two lighter nuclei?

    (1) titration   (3) electrolysis

    (2) fission   (4) neutralization   28 _____

29 Which process converts mass into energy?

  (1) distillation of ethanol
  (2) filtration of a mixture
  (3) fusion of hydrogen atoms
  (4) ionization of cesium atoms   29 _____

30 Which radioisotope is used to determine the age of once-living organisms?

  (1) carbon-14   (3) iodine-131
  (2) cobalt-60   (4) uranium-238   30 _____

PART B–1

Answer all questions in this part.

Directions (31–50): For *each* statement or question, write in the answer space the *number* of the word or expression that, of those given, best completes the statement or answers the question. Some questions may require the use of the *2011 Edition Reference Tables for Physical Setting/Chemistry.*

31 Which electron configuration represents the electrons in an atom of calcium in an excited state?

(1) 2–8–8   (3) 2–7–8–1
(2) 2–8–8–2   (4) 2–7–8–3   31 _____

32 The table below gives the atomic mass and the abundance of the two naturally occurring isotopes of bromine.

Naturally Occurring Isotopes of Bromine

Isotopes	Atomic Mass (u)	Natural Abundance (%)
Br-79	78.92	50.69
Br-81	80.92	49.31

Which numerical setup can be used to calculate the atomic mass of the element bromine?

(1) (78.92 u)(50.69) + (80.92 u)(49.31)
(2) (78.92 u)(49.31) + (80.92 u)(50.69)
(3) (78.92 u)(0.5069) + (80.92 u)(0.4931)
(4) (78.92 u)(0.4931) + (80.92 u)(0.5069)   32 _____

33 Given the formulas of two substances:

$$\ddot{O}{=}\ddot{O} \qquad \ddot{O}{=}\ddot{O}{\diagdown}\underset{\cdot\cdot}{\overset{\cdot\cdot}{O}}{:}$$

These diagrams represent substances that have

(1) the same molecular structure and the same physical properties
(2) the same molecular structure and different physical properties
(3) different molecular structures and the same physical properties
(4) different molecular structures and different physical properties   33 _____

34 What is the chemical formula of titanium(II) oxide?

(1) TiO   (3) TiO_2
(2) Ti_2O (4) Ti_2O_3 34 _____

35 Which equation shows conservation of mass and energy for a reaction at 101.3 kPa and 298 K?

(1) $2H_2(g) + O_2(g) \rightarrow 2H_2O(g) + 483.6$ kJ
(2) $2H_2(g) + O_2(g) \rightarrow 2H_2O(\ell) + 285.8$ kJ
(3) $H_2(g) + O_2(g) \rightarrow H_2O(g) + 483.6$ kJ
(4) $H_2(g) + O_2(g) \rightarrow H_2O(\ell) + 285.8$ kJ   35 _____

36 Which two particle diagrams each represent a sample of one substance?

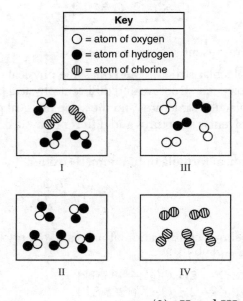

(1) I and II	(3) II and III	
(2) I and III	(4) II and IV	36 _____

37 According to Table *G*, which substance forms an unsaturated solution when 80. grams of the substance are stirred into 100. grams of H_2O at 10.°C?

(1) KNO_3	(3) NH_3	
(2) KI	(4) NaCl	37 _____

38 What is the concentration of AgCl in an aqueous solution that contains 1.2×10^{-3} gram of AgCl in 800. grams of the solution?

(1) 1.2 ppm	(3) 7.2 ppm	
(2) 1.5 ppm	(4) 9.6 ppm	38 _____

39 A sample of gas is in a rigid cylinder with a movable piston. The pressure of the gas is kept constant. If the Kelvin temperature of the gas is doubled, the volume of the gas is

(1) halved	(3) tripled	
(2) doubled	(4) unchanged	39 _____

40 What is the amount of heat required to completely melt a 200.-gram sample of $H_2O(s)$ at STP?

(1) 334 J   (3) 66 800 J

(2) 836 J   (4) 452 000 J   40 _____

41 As a 15.1-gram sample of a metal absorbs 48.75 J of heat, its temperature increases 25.0 K. What is the specific heat capacity of the metal?

(1) 0.129 J/g·K   (3) 3.23 J/g·K

(2) 1.95 J/g·K   (4) 7.74 J/g·K   41 _____

42 Given the formula:

$$\begin{array}{ccccc} & H & H & O & \\ & | & | & \| & \\ H- & C & - C & - C & -H \\ & | & | & & \\ & H & H & & \end{array}$$

What is a chemical name of this compound?

(1) propane   (3) propanol

(2) propanal   (4) propanone   42 _____

43 Given the equation representing a reaction:

$$\begin{array}{ccccccc} H & & & & H & & \\ | & & & & | & & \\ H-C-H & + & Cl-Cl & \rightarrow & H-C-H & + & H-Cl \\ | & & & & | & & \\ H & & & & Cl & & \end{array}$$

Which type of reaction is represented by this equation?

(1) addition   (3) polymerization

(2) esterification   (4) substitution   43 _____

44 Atoms of which element react spontaneously with $Mg^{2+}(aq)$?

(1) chromium   (3) iron

(2) barium   (4) zinc   44 _____

45 In a titration, 5.0 mL of a 2.0 M NaOH(aq) solution exactly neutralizes 10.0 mL of an HCl(aq) solution. What is the concentration
of the HCl(aq) solution?

(1) 1.0 M   (3) 10. M
(2) 2.0 M   (4) 20. M   45 _____

46 Given the equation representing a reaction at equilibrium:

$$NH_3(g) + H_2O(\ell) \rightleftharpoons NH_4^+(aq) + OH^-(aq)$$

If an acid is defined as an H^+ donor, what is the acid in the forward
reaction?

(1) OH^-(aq)   (3) $NH_3(g)$
(2) $H_2O(\ell)$ (4) NH_4^+(aq)   46 _____

47 Compared to a solution with a pH value of 7, a solution with a
thousand times greater hydronium ion concentration has a pH
value of

(1) 10   (3) 3
(2) 7   (4) 4   47 _____

48 Which Lewis electron-dot diagram represents the bonding in
potassium iodide?

$K^+ \left[:\ddot{I}: \right]^-$ $K:\ddot{I}:$

(1)   (3)

$\left[:\ddot{K}: \right]^- I^+$ $:\ddot{K}:I$

(2)   (4)   48 _____

49 The table below shows properties of two compounds at standard pressure.

Selected Properties of Two Compounds

Compound	Melting Point (°C)	Boiling Point (°C)	Electrical Conductivity
1	775	1935	good as a liquid or in an aqueous solution
2	−112.1	46	poor as a liquid

Which statement classifies the two compounds?

(1) Both compounds are ionic.
(2) Both compounds are molecular.
(3) Compound 1 is ionic, and compound 2 is molecular.
(4) Compound 1 is molecular, and compound 2 is ionic.   49 _____

50 Given two balanced equations, each representing a reaction:

Equation 1:

$$^{226}_{88}Ra \rightarrow {}^{222}_{86}Rn + {}^{4}_{2}He + 4.8 \times 10^8 \text{ kJ}$$

Equation 2:

$$C_3H_8 + 5O_2 \rightarrow 3CO_2 + 4H_2O + 2.2 \times 10^3 \text{ kJ}$$

Which statement compares the energy terms in these two equations?

(1) Equation 1 shows 2.2×10^5 times more energy being absorbed.
(2) Equation 2 shows 2.2×10^5 times more energy being absorbed.
(3) Equation 1 shows 2.2×10^5 times more energy being released.
(4) Equation 2 shows 2.2×10^5 times more energy being released.   50 _____

PART B–2

Answer all questions in this part.

Directions (51–65): Record your answers on the answer sheet provided. Some questions may require the use of the *2011 Edition Reference Tables for Physical Setting/Chemistry*.

Base your answers to questions 51 and 52 on the information below and on your knowledge of chemistry.

The formula below represents a molecule of butanamide.

$$
\begin{array}{ccccc}
\text{H} & \text{H} & \text{H} & \text{O} & \text{H} \\
| & | & | & \| & / \\
\text{H}-\text{C}-\text{C}-\text{C}-\text{C}-\text{N} & & & & \\
| & | & | & & \backslash \\
\text{H} & \text{H} & \text{H} & & \text{H}
\end{array}
$$

51 State the type of chemical bond between a hydrogen atom and the nitrogen atom in the molecule.   [1]

52 Explain, in terms of charge distribution, why a molecule of butanamide is polar.   [1]

Base your answers to questions 53 and 54 on the information below and on your knowledge of chemistry.

An equilibrium system in a sealed, rigid container is represented by the equation below.

$$CO(g) + H_2O(g) \rightleftharpoons CO_2(g) + H_2(g)$$

53 Compare the rate of the forward reaction to the rate of the reverse reaction at equilibrium.   [1]

54 State the effect on the concentrations of $H_2O(g)$ and $CO_2(g)$ when more $H_2(g)$ is added to the system.   [1]

Base your answers to questions 55 through 58 on the information below and on your knowledge of chemistry.

The table below contains selected information about chlorine and two compounds containing chlorine. One piece of information is missing for each of the substances in the table.

Chlorine and Two Compounds Containing Chlorine

Name	Formula	Molar Mass (g/mol)	Phase at STP
chlorine	Cl_2	71	?
calcium chloride	$CaCl_2$	?	solid
1,2-dichloroethene	?	97	liquid

55 Identify the phase of the chlorine at STP.   [1]

56 Determine the molar mass for calcium chloride.   [1]

57 The liquid compound has an empirical formula of CHCl. Write the molecular formula for this compound.   [1]

58 Explain, in terms of electrons, why the compound containing calcium and chlorine is classified as an ionic compound.   [1]

Base your answers to questions 59 and 60 on the information below and on your knowledge of chemistry.

The equation below represents the reaction between 2-methylpropene and hydrogen chloride gas.

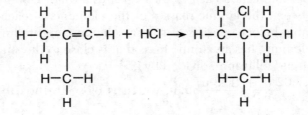

59 Explain, in terms of chemical bonds, why the hydrocarbon is unsaturated.   [1]

60 Identify the class of organic compounds to which the product belongs.   [1]

Base your answers to questions 61 through 65 on the information below and on your knowledge of chemistry.

Many scientists made observations of the elements that led to the modern Periodic Table. In 1829, Dobereiner found groups of three elements that have similar properties and called each of these groups a triad. Dobereiner noticed a relationship between the atomic masses of the elements in each triad. Triad 1, shown in the table below, consists of sulfur, selenium, and tellurium. The middle element, selenium, has an atomic mass that is close to the sum of the atomic masses of sulfur and tellurium, divided by 2.

For example: $\dfrac{32 \text{ u} + 128 \text{ u}}{2}$ 80. u, which is close to the 79 u value in the table.

The other triads shown in the table below demonstrate the same mathematical relationship.

Dobereiner's Triads

Triad	Triad	Dobereiner's Atomic Masses (u)
1	sulfur selenium tellurium	32 79 128
2	calcium strontium barium	40. 88 137
3	chlorine bromine iodine	35.5 80. 127
4	lithium sodium potassium	7 23 39

61 Identify the triad that contains a metalloid. [1]

62 Explain, in terms of electrons, why the elements in triad 2 have similar chemical properties. [1]

63 State the trend in first ionization energy as the elements in triad 3 are considered in order of increasing atomic number. [1]

64 Compare the volume of a 100.-gram sample of the first element in triad 4 to the volume of a 100.-gram sample of the third element in triad 4 when both samples are at room temperature.   [1]

65 Show a numerical setup that demonstrates Dobereiner's mathematical relationship for triad 2.   [1]

PART C

Answer all questions in this part.

Directions (66–85): Record your answers on the answer sheet provided. Some questions may require the use of the *2011 Edition Reference Tables for Physical Setting/Chemistry*.

Base your answers to questions 66 through 68 on the information below and on your knowledge of chemistry.

Wood is mainly cellulose, a polymer produced by plants. One use of wood is as a fuel in campfires, fireplaces, and wood furnaces. The molecules of cellulose are long chains of repeating units. Each unit of the chain can be represented as $C_6H_{10}O_5$. The balanced equation below represents a reaction that occurs when $C_6H_{10}O_5$ is burned in air.

$$C_6H_{10}O_5 + 6O_2 \rightarrow 6CO_2 + 5H_2O + \text{heat}$$

66 State evidence from the equation that this reaction is exothermic. [1]

67 Explain, in terms of substances in the reaction, why the equation represents a chemical change. [1]

68 Show a numerical setup for calculating the percent composition by mass of carbon in $C_6H_{10}O_5$ (gram-formula mass = 162.1 g/mol). [1]

Base your answers to questions 69 through 72 on the information below and on your knowledge of chemistry.

Millions of tons of ammonia are produced each year for use as fertilizer to increase food production. Most of the hydrogen needed to produce ammonia comes from methane gas reacting with steam. This reaction, which occurs in a container under controlled conditions, is shown below in unbalanced equation 1.
Equation 1:

$$CH_4(g) + H_2O(g) + energy \rightarrow CO(g) + H_2(g)$$

The reaction that produces ammonia is represented by balanced equation 2, shown below. A catalyst can be used to increase the rate of the reaction.
Equation 2:

$$N_2(g) + 3H_2(g) \rightarrow 2NH_3(g) + energy$$

A potential energy diagram for equation 2 is shown below.

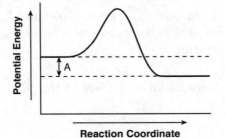

69 Balance equation 1 *on your answer sheet*, using the smallest whole-number coefficients. [1]

70 Explain, in terms of collision theory, why an increase in temperature increases the rate of reaction between methane gas and steam. [1]

71 State what is represented by interval A on the potential energy diagram. [1]

72 Determine the number of moles of hydrogen gas required to react completely with 50.0 moles of nitrogen gas in the production of ammonia. [1]

Base your answers to questions 73 through 77 on the information below and on your knowledge of chemistry.

Diethyl ether is used as a laboratory and industrial solvent. The boiling point of diethyl ether at standard pressure is 34.6°C. The equation below represents a reaction that produces diethyl ether.

73 Identify the element in diethyl ether that allows it to be classified as an organic compound.   [1]

74 State the number of electrons shared between the carbon atoms in one molecule of the organic reactant.   [1]

75 State why the reaction is classified as a synthesis reaction.   [1]

76 Explain, in terms of the strength of intermolecular forces, why the boiling point of diethyl ether at standard pressure is *lower* than the boiling point of water at standard pressure.   [1]

77 Draw a structural formula for an isomer of the product that has the same functional group.   [1]

Base your answers to questions 78 through 81 on the information below and on your knowledge of chemistry.

The electrolysis of brine, a concentrated aqueous sodium chloride solution, produces three important industrial chemicals: chlorine gas, hydrogen gas, and sodium hydroxide. The diagram and equation below represent a brine electrolysis cell. Before the battery is connected, the pH value of the brine solution is 7.0.

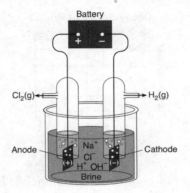

$$2NaCl(aq) + 2H_2O(\ell) \rightarrow Cl_2(g) + H_2(g) + 2NaOH(aq)$$

78 Explain, in terms of energy, why this cell is an electrolytic cell. [1]

79 Explain, in terms of ions, why the aqueous solution in the cell conducts an electric current. [1]

80 State the oxidation number of oxygen in the aqueous product. [1]

81 Compare the pH value of the solution before the battery is connected to the pH value of the solution after the cell operates for 20 minutes. [1]

Base your answers to questions 82 through 85 on the information below and on your knowledge of chemistry.

The isotope Rn-222 is produced by the decay of uranium in Earth's crust. Some of this isotope leaks into basements of homes in areas where the ground is more porous. An atom of Rn-222 decays to an atom of Pb-206 through a series of steps as shown on the graph below.

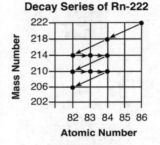

Decay Series of Rn-222

82 Determine the number of neutrons in an atom of Pb-214.   [1]

83 Complete the nuclear equation *on your answer sheet* for the decay of Po-218 by writing a notation for the missing product.   [1]

84 Determine the fraction of an original sample of Rn-222 that remains unchanged after 7.646 days.   [1]

85 Explain, in terms of elements, why the decay of Bi-210 is considered a transmutation.   [1]

Answer Sheet
August 2017
Chemistry—Physical Setting

PART B–2

51 _____

52 _____

_____

_____

53 _____

_____

_____

54 $H_2O(g)$: _____

$CO_2(g)$: _____

55 _____

56 _____ **g/mol**

57 _____

58 _____

_____

_____

59 _____

60 _____

61 _____

62 _____

63 _____

64 _____

65

PART C

66 _____

_____

_____

67 _____

_____

_____

68

69 ___ $CH_4(g)$ + ___ $H_2O(g)$ + energy → ___ $CO(g)$ + ___ $H_2(g)$

70 _____

_____

_____

71 _____

_____

_____

72 _____ **mol**

73 _____

74 _____

75 _____

_____

_____

76 _____

_____

_____

77

78 _____

_____

_____

79 _____

_____

_____

80 _____

81 _____

_____

_____

82 _____

83 $^{218}_{84}PO \rightarrow ^{214}_{82}PB +$ _____

84 _____

85 _____

_____

_____

Answers
August 2017
Chemistry—Physical Setting

Answer Key

PART A

1. 3	7. 4	13. 1	19. 1	25. 3
2. 2	8. 1	14. 4	20. 4	26. 1
3. 2	9. 3	15. 4	21. 2	27. 4
4. 4	10. 1	16. 3	22. 2	28. 2
5. 1	11. 3	17. 4	23. 2	29. 3
6. 4	12. 4	18. 1	24. 1	30. 1

PART B–1

31. 4	35. 1	39. 2	43. 4	47. 4
32. 3	36. 4	40. 3	44. 2	48. 1
33. 4	37. 2	41. 1	45. 1	49. 3
34. 1	38. 2	42. 2	46. 2	50. 3

PART B–2 and **PART C**. *See* **Answers Explained**.

Answers Explained

PART A

1. **3** The basic structure of an atom is a positively charged nucleus surrounded by a negatively charged electron cloud. The nucleus contains protons, which have a positive charge. Reference Table O provides a handy summary of the names and charges of protons, neutrons, and electrons.

2. **2** Atoms are neutral in charge, since the number of positively charged protons is equal to the number of negatively charged electrons. An atom with 20 protons would have 20 electrons.

3. **2** Reference Table O lists the symbols for all of the particles mentioned in this problem, but you do need to remember that an electron is identical in mass and charge to a beta particle. Particle notations for the neutron and the proton are shown below:

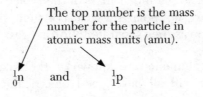

The top number is the mass number for the particle in atomic mass units (amu).

^1_0n and ^1_1p

Protons and neutrons both have a mass of approximately 1 amu.

WRONG CHOICES EXPLAINED:

(1), (4) An electron has the same mass and charge as a beta particle. Looking at Reference Table O, we can see that the mass of an electron is 0.

In fact, the electron does have a mass of $\dfrac{1}{1840}$ amu, but this is negligible compared with the mass of a proton or a neutron.

(3) Reference Table O shows that the mass of an alpha particle is 4 amu.

4. **4** When a solid becomes a gas, the particles break free of their tightly packed lattice arrangement and become randomly positioned with respect to each other. This represents a dramatic change in particle arrangement! This change in particle arrangement is shown in the diagram below:

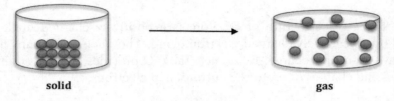

solid   **gas**

WRONG CHOICES EXPLAINED:

(1) The bonds between carbon and oxygen atoms in CO_2 molecules are non-polar covalent bonds. The bonds within CO_2 molecules do not change when the molecules move farther apart.

(2) The gram-formula mass (also known as the molar mass) of CO_2 is 44.0 g/mol. This is calculated by adding the individual molar masses of the atoms in the compound. Molecules of CO_2 have the same composition in the solid phase as they do in the gas phase, so the molar mass of the molecules does not change.

(3) CO_2 molecules are nonpolar by virtue of their symmetry. Molecules of CO_2 do not change shape as they separate to form a gas.

5. **1** The Group 2 elements are known as the Alkaline Earth Metals. As metals, these elements will all have good electrical conductivity. Reference Table *S* lists the values for the electronegativity of most elements. The electro-negativity values for Group 2 elements range from a high of 1.6 for beryllium to a low of 0.9 for radium. All of these values are *below* 1.7.

6. **4** In the element notation used in this question, the numbers shown are the mass numbers of the two isotopes. Different isotopes of any element will always have different mass numbers because they have different numbers of neutrons. Remember that the mass number is equal to the number of neutrons in an atom added to the number of protons.

WRONG CHOICES EXPLAINED:

(1) Six electrons are present in every atom of carbon. The number of electrons always equals the number of protons in all atoms since atoms have no overall charge.

(2), (3) All atoms of the same element have the same number of protons, which is equal to the element's atomic number. An element is defined by its atomic number. Six protons are present in every atom of carbon since carbon's atomic number is 6.

7. **4** The Group 18 elements are known as the noble gases. Noble gases are considered to be the most stable elements on the Periodic Table because they usually do not react with other elements. These gases rarely form bonds with other elements since they already possess the stable octet of 8 valence electrons.

8. **1** According to the Periodic Table in your Reference Tables, magnesium atoms have a ground-state configuration of 2-8-2. These numbers represent the number of electrons found in the first, second, and third electron shells, respectively. The 2 third-shell electrons are lost when magnesium loses two electrons, leaving the resulting magnesium ion with an electron configuration of 2-8. Since the magnesium ion still has 12 positively charged protons but now only has 10 negatively charged electrons, the magnesium ions have a positive charge. In addition, the ion is smaller than the original atom because the third electron shell is no longer occupied.

9. **3** You must know the definition of electronegativity to answer this question correctly. Electronegativity is a measure of how strongly atoms attract bonding electrons to themselves. Electronegativity values for most elements are listed in Reference Table S. Chlorine's electronegativity of 3.2 is higher than that of any of the other elements given as answer choices for this question. A familiarity with periodic trends can also be helpful when answering this question, namely knowing that electronegativity values increase from left to right across any period of the Periodic Table.

10. **1** Elements are the simplest type of matter. Elements are composed of atoms, which can only be broken down by nuclear change, not by chemical change.

WRONG CHOICES EXPLAINED:
(2), (3) A solution is a homogeneous mixture. All mixtures can be physically separated.
(4) Compounds consist of two or more elements that are chemically bonded together. Compounds can be broken down into their constituent elements by chemical reactions (chemical change).

11. **3** Sodium (Na) is a Group 1 element. The first column in Reference Table F shows that all compounds containing sodium ions are soluble.

WRONG CHOICES EXPLAINED:

(1) Silver (Ag) ions are listed in the second column of Reference Table F as an exception to the solubility of halide ions, such as the Cl^- ion. Think of the second column as "not soluble," which is what it means to be an "exception to solubility."

(2) Carbonates (compounds containing the CO_3^{2-} ion) are listed in the third column of Reference Table F as an example of ions that form *insoluble* compounds in water. The calcium ion is not listed in the last column as an exception.

(4) Most compounds that contain the sulfate ion, SO_4^{2-}, are soluble in water. Notice that sulfates are listed in the first column of Reference Table F. However, the strontium ion, Sr^{2+}, is listed in the second column as an exception to the solubility of sulfates. In other words, $SrSO_4$ is *not* soluble.

12. **4** This reaction shows two hydrogen atoms bonding together to form one H_2 molecule. The formation of chemical bonds between elements always causes a *release* of potential energy since elements bond to become more stable.

13. **1** Air is a mixture of several gases, including N_2, O_2, and CO_2.

WRONG CHOICES EXPLAINED:

(2), (4) Ammonia (NH_3) and water (H_2O) are compounds. Compounds are pure substances, not mixtures.

(3) Manganese (Mn) is an element on the Periodic Table (atomic number = 25). Elements are pure substances, not mixtures.

14. **4** Paper chromatography works because the different components, or pigments, in a mixture of colored dyes have different levels of attraction to paper. As the dye mixture advances up the paper, pigment compounds that are more attracted to the paper will absorb first, while compounds with less attraction will move farther up the paper before being absorbed. The attraction between dye compounds and paper is caused by molecular polarity (the attraction of opposite partial charges on different molecules).

15. **4** Dissolved ions raise the boiling point of a solution. This is the colligative property known as boiling point elevation. The more ions that are dissolved, the higher the boiling point. As more $NaNO_3(s)$ is dissolved in the solution, the boiling point will increase.

16. **3** Thermal energy is created by the motion of atoms and molecules within an object.

WRONG CHOICES EXPLAINED:
(1) Combustion is a type of chemical reaction. It is not a form of energy.
(2) The term "exothermic" describes a chemical reaction that releases energy.
(4) "Electrolytic" is a type of electrochemical cell that utilizes electrical energy to produce a chemical change.

17. **4** One of the principles of the kinetic molecular theory of gases is that gas particles move in random, constant, straight-line motion. This allows gas particles to expand to fill the space that they occupy. Other principles of the kinetic molecular theory of gases state that an ideal gas consists of particles that:

- have mass
- have no attractions or repulsions for each other
- are separated by distances much greater than their size
- undergo perfectly elastic collisions in which the total energy of the system remains the same before and after collisions

18. **1** Temperature is defined as a measure of the average kinetic energy of the particles in any substance.

19. **1** Real gases behave more like ideal gases when temperature is high and pressure is low. These conditions minimize any possible attractions between the gas particles, allowing the gas to adhere more closely to the principles of the kinetic molecular theory of gases. (A good way to remember this is to think of summertime, when the temperature is high and the pressure is low.)

20. **4** The basic idea behind collision theory is that particles will only react if they have sufficient energy and proper orientation.

21. **2** Equal volumes of gas at the same conditions of temperature and pressure contain the same number of particles. This relationship is known as Avogadro's Law. For example, the molar volume of a gas at STP equals 22.4 L/mol no matter what type of gas you have.

22. **2** Energy is required to overcome attractions between particles in order to increase spacing between them. Conversely, energy is released when particles that are attracted are able to be closer together. A physical change from the liquid

phase to the solid phase puts particles much closer together, corresponding to a release in potential energy. Consider that to make solid ice, liquid water must be put in the freezer, where energy is taken out of the water.

23. **2** All of the equations shown represent equilibrium systems, but only answer choice (2) shows chemical equilibrium in which forward and reverse chemical reactions are happening at equal rates. The other three answer choices all show changes in phase, and these are physical changes, not chemical changes.

24. **1** Entropy is defined as the measure of randomness, or disorder, in a system.

25. **3** All chemical reactions conserve mass, energy, and charge. If one mole of electrons is lost as $Ca(s)$ oxidizes, then one mole of electrons must be gained by the substance that is reducing, in this case $O_2(g)$.

26. **1** The mnemonic device "An Ox, Red Cat" is helpful for remembering that oxidation (loss of electrons) takes place at the anode, while reduction (gain of electrons) happens at the cathode.

27. **4** Three classes of compounds are electrolytes: acids, bases, and salts. Electrolytes are compounds that dissociate into ions in aqueous solution and, because of these freely moving ions, enable the flow of electricity through the solution. H_2SO_4 is sulfuric acid (check Reference Table K for a list of common acids). It dissociates into H^+ and SO_4^{-2} ions in aqueous solution. All of the other answer choices are nonelectrolytes, meaning that they are covalently bonded and would not dissociate.

28. **2** Fission is the splitting of a large atomic nucleus (usually U-235) into two smaller nuclei. The fission reaction is initiated by the bombardment of the target nucleus with a neutron, and the resulting emission of neutrons as products makes a chain reaction possible. An example of a fission reaction is shown below:

$$^{235}_{92}U + ^{1}_{0}n \rightarrow ^{94}_{38}Sr + ^{140}_{54}Xe + 2^{1}_{0}n$$

WRONG CHOICES EXPLAINED:

(1) Titration is a laboratory process in which the concentration of an unknown solution can be determined by the careful addition of a standard solution with a known concentration. Titration often involves the neutralization of an acid with a base (or vice versa).

(3) Electrolysis is a redox reaction in which electricity is used to force a non-spontaneous chemical change to happen.

(4) Neutralization reactions occur between acids and bases when H^+ ions react with OH^- ions to form water.

29. **3** Nuclear changes produce energy by converting mass into energy. Of the answer choices listed, fusion is the only nuclear reaction.

30. **1** Since living organisms are carbon-based organisms, the radioisotope C-14 is helpful in determining the age of ancient remains from previously living organisms. The process known as radiocarbon dating involves the analysis of the amount of carbon-14 remaining in the sample.

PART B–1

31. **4** An atom of calcium will always have 20 electrons. The Periodic Table in the Reference Tables shows that the ground-state electron configuration of a calcium atom is 2-8-8-2. A calcium atom would be in an excited state if its configuration were 2-7-8-3 because this configuration shows that a second-shell electron has moved to the fourth shell.

WRONG CHOICES EXPLAINED:
(1) There are only 18 electrons shown in this configuration. This is the ground-state configuration for the calcium *ion*, Ca^{+2}, but not for a calcium *atom*.

(2) This is the ground-state configuration for a calcium atom. The question asks for an *excited*-state configuration.

(3) This configuration is short 2 electrons. All calcium atoms have 20 electrons.

32. **3** The atomic mass of an element is found by taking the weighted average of all naturally occurring isotopes of that element. The natural abundance percentages of each isotope are first converted to decimal fractions, and these are then multiplied by the atomic mass of the isotope. Adding the contributions of each isotope together yields the correct atomic mass for that element.

WRONG CHOICES EXPLAINED:
(1) Although the correct percentages are multiplied by each atomic mass, the final sum would have to be divided by 100.

(2) This numerical setup multiplies each isotope's atomic mass by the wrong percentage. Another error is that, if the actual percentage values are used without converting them to decimal fractions, the final step must be to divide by 100, since a percent is literally "parts per hundred."

(4) This numerical setup shows the percentages converted to decimal fractions, but they are then multiplied by the opposite isotope's atomic mass.

33. **4** The substances shown are two different allotropes of the element oxygen. Allotropes are different naturally occurring structures of the same element. Since the molecular structures are different, the molecules would exhibit different physical properties, such as melting point, boiling point, and density.

34. **1** Titanium(II) oxide is an ionic compound made from the titanium(II) ion, Ti^{+2}, and the oxide ion, O^{-2}. Charges for these two ions can be found on the Periodic Table. Since compounds have no overall charge, one titanium(II) ion will cancel the -2 charge of the oxide ion.

WRONG CHOICES EXPLAINED:
(2) This would be the formula for titanium(I) oxide. In this compound, it takes two Ti^{+1} ions to cancel the -2 charge of the oxide ion.

(3) This is the formula for titanium(IV) oxide. One Ti^{+4} ion would cancel the charge of two oxide ions, each having a charge of -2.

(4) This is the formula for titanium(III) oxide. Two Ti^{+3} ions will cancel the charge of three O^{-2} ions.

35. **1** You are not expected to memorize the heat of reaction for chemical reactions, so the only way to know which answer choice is correct is to consult Reference Table *I*, which lists the heat of reaction for some common chemical reactions at 101.3 kPa and 298 K. The reaction in answer choice (1) is identical to one of the equations listed in this table. The ΔH for this reaction is listed as -483.6 kJ. Since the heat of reaction has a negative value, the reaction is exothermic, and the heat term is written as a product in the chemical equation.

WRONG CHOICES EXPLAINED:
(2) This equation is balanced for mass, but the heat of reaction is incorrect. The correct heat of reaction for this reaction is shown on Reference Table *I*.

(3) This equation is not balanced for mass. Two oxygen atoms appear on the reactant side of this equation, but only one oxygen atom appears on the product side.

(4) This equation is neither balanced for mass nor is it balanced for energy.

36. **4** A pure substance is made from just one type of particle. Diagram II shows just one type of molecule, and Diagram IV shows just one type of diatomic element. Diagrams I and III are incorrect because they represent mixtures of more than one substance.

37. **2** The lines on Reference Table *G* represent lines of solution saturation. A saturated solution is holding all of the solute possible at a given temperature. An unsaturated solution is not fully saturated, meaning that more solute could dissolve. On Reference Table *G*, look carefully at the intersection between 10°C and 80 grams of solute (in 100 grams of water). This intersection is *above* the saturation curves for KNO_3, NH_3, and NaCl. Adding 80 grams of any of these compounds to 100 grams of water would create *supersaturated* solutions. Only the saturation curve for KI is above the intersection between 10°C and 80 grams of solute. Since 80 grams of KI would not saturate a solution made in 100 grams of water at 10°C, the solution would be *unsaturated*.

38. **2** Look at Reference Table *T* to find the formula for parts per million:

$$\text{parts per million} = \frac{\text{mass of solute}}{\text{mass of solution}} \times 1{,}000{,}000$$

$$\text{parts per million} = \frac{0.0012 \text{ g}}{800. \text{ g}} \times 1{,}000{,}000 = 1.5 \text{ ppm}$$

39. **2** The Combined Gas Law found on Reference Table *T* is useful for thinking about this problem. Since the pressure is kept constant, we can reduce the Combined Gas Law to the following expression, which is known as Charles's Law:

$$\frac{V_1}{T_1} = \frac{V_2}{T_2}$$

There is a direct relationship between the volume of a gas and the Kelvin temperature of a gas at constant pressure. If one doubles, the other must also double. For example, in the illustration below, if the Kelvin temperature increases, the gas particles will strike the piston more often and with more force, and the piston will push outward until the pressure inside and outside becomes equal.

40. **3** The last page of the Reference Tables, Reference Table T, lists the formula to find the heat of fusion, which is the heat lost or gained by a substance as it freezes or melts, respectively.

$$q = mH_f$$

where q = heat, m = mass, and H_f = heat of fusion

The mass is given as 200. grams, and the heat of fusion for liquid water, found on Reference Table B, is 334 J/g. Plug these values into the heat of fusion formula and solve:

$$q = (200.\ g)(334\ J/g) = 66800\ J$$

41. **1** To solve for the specific heat capacity, use the following equation found on Reference Table T:

$$q = mC\Delta T$$

where q = heat, m = mass,
C = specific heat capacity,
and ΔT = change in temperature

Rearrange this equation to solve for C:

$$C = \frac{q}{m\Delta T}$$

Now just plug in the values given in the problem:

$$C = \frac{48.75\ J}{(15.1\ g)(25.0\ k)} = 0.129\ J/g{\cdot}K$$

42. **2** Refer to Reference Table R for all questions concerning organic functional groups. The structure shown in this question is actually found as an example directly on this table! The compound is definitely an aldehyde since it has the aldehyde functional group shown below:

Since there are three carbon atoms in the molecule, the correct prefix from Reference Table *P* would be "prop-." To finish the name, use the aldehyde suffix "-al" as shown in the example on Reference Table *R*. Putting these pieces together, the compound is correctly named propanal.

43. **4** This is an organic reaction known as a substitution reaction because one of the chlorine atoms substitutes for one of the hydrogen atoms on the hydrocarbon.

WRONG CHOICES EXPLAINED:

(1) Addition reactions require a hydrocarbon reactant with a double or triple carbon–carbon bond. The complete diatomic molecule would "add" to the double bond: in other words, the product would contain *both* chlorine atoms, not just one.

(2) Esterification reactions produce esters. The product shown does not contain the ester functional group.

(3) Polymerization reactions create long chains from many simple identical molecules, called monomers. The product in this reaction is a small, single carbon molecule.

44. **2** Reference Table *J* lists metals from the most active metals at the top to the least active metals at the bottom. Pure metal atoms never gain electrons, but metal ions can gain back the electrons they previously lost in the presence of a more active metal. Mg^{+2} ions will gain electrons back in the presence of barium, which is higher on Reference Table *J* and is therefore a more active metal.

45. **1** Use the titration formula found on Reference Table *T*:

$$M_A V_A = M_B V_B$$

where M_A = molarity of H^+,
V_A = volume of acid,
M_B = molarity of OH^-,
and V_B = volume of base

Substitute the values from this problem:

$$(M_A)(10.0 \text{ mL}) = (2.0 \text{ M})(5.0 \text{ mL})$$

$$M_A = \frac{(2.0 \text{ M})(5.0 \text{ mL})}{10.0 \text{ mL}} = 1.0 \text{ M}$$

46. **2** Compare the reactant formulas to the product formulas. $NH_3(g)$ becomes $NH_4^+(aq)$ by gaining a hydrogen ion (H^+) from $H_2O(\ell)$. In other words, $H_2O(\ell)$ "donates" a hydrogen ion to $NH_3(g)$ to form NH_4^+.

47. **4** pH is a function of the base 10 exponent of the hydronium ion concentration in an aqueous solution. As an example, a solution with a hydronium concentration of 0.001 M $(= 10^{-3}$ M) has a pH of $-(-3)$, in other words a pH of 3. Likewise, a solution with a hydronium concentration of 0.0001 M $(=10^{-4}$ M) has a pH of 4. Each number on the pH scale differs in hydronium ion concentration by a factor of 10 from the next whole number higher and lower on the scale. In this problem, a sample that has one thousand times *greater* hydronium ions would differ by *three* whole numbers from the original pH of 7. Having more hydronium ions means that the sample is more acidic than the original solution and would have a lower pH value. Putting this information together, the pH of the new sample would be $7 - 3 = 4$.

48. **1** Lewis diagrams for ionic compounds show both *ion charges* and *electron transfer*, with all eight valence electrons surrounding the nonmetal, in this case iodine.

WRONG CHOICES EXPLAINED:
(2) The charges and electron positions in this answer choice are incorrect. Metal atoms always lose electrons when they form ionic bonds. Metal ions always have a positive charge. Nonmetal atoms always gain electrons from metal atoms when ionic bonds form. The electrons gained complete the stable octet of valence electrons around the nonmetal atom.

(3), (4) These answer choices show the formation of a shared pair of electrons (covalent bond) between potassium (a metal) and iodine (a nonmetal). Bonds between metal and nonmetal atoms are always ionic bonds, not covalent.

49. **3** Compound 1 has a high melting point and a high boiling point, both of which are characteristic of ionic compounds. In addition, ionic compounds will conduct electricity *only* when the compound is melted or dissolved and the ions are free to move. The extremely low melting point and low boiling point of Compound 2 are characteristic of molecular compounds, which are held together with intermolecular forces. Intermolecular forces are much weaker than ionic bonds, so molecular compounds melt and boil at temperatures that are far lower than those of ionic compounds.

50. **3** Since the heat term is on the product side of both reactions, heat is released in both reactions. This eliminates both answer choices (1) and (2). The heat term in Equation 1 is on the order of 10^5 times larger than the heat term in Equation 2.

PART B–2

[All questions in Part B–2 are worth 1 point.]

51. Since both hydrogen and nitrogen are nonmetals, they will form a covalent bond. Since these elements have different electronegativities, they will form a **polar covalent bond** in which electrons are shared unequally.
   (NOTE: **Covalent bond** was also accepted as a correct response to this question.)

52. The oxygen and nitrogen atoms on the end of the butanamide molecule have a far greater electronegativity than the carbon and hydrogen atoms on the other end of the molecule. Electrons will be drawn disproportionately toward the oxygen/nitrogen end of the molecule, giving the overall molecule an **asymmetrical (or unequal) distribution of charge**. Butanamide is a polar molecule because the electrons are not distributed evenly throughout the molecule but are drawn more to one end than the other.

53. When chemical equilibrium is established, **the rate of the forward reaction equals the rate of the reverse reaction**. This is the definition of chemical equilibrium.

54. The addition of $H_2(g)$ is a stress on this equilibrium system, and according to Le Châtelier's principle, systems at equilibrium will "shift" to reduce the stress and restore equilibrium. By "shifting to the left," the extra $H_2(g)$ that was added will be used up. "Shifting to the left" means that the substances on the left of the reaction arrow are produced while the substances on the right of the reaction arrow are consumed. When a new equilibrium is established, **the concentration of $H_2O(g)$ will be higher while the concentration of $CO_2(g)$ will be lower**.

55. Recall that "STP" means "Standard Temperature and Pressure." According to Reference Table A, standard temperature is defined as 273 K. Use Reference Table S to find the melting point and the boiling point information for chlorine. Since chlorine's boiling point is 239 K, this element will be a **gas** at standard temperature.

56. The molar mass of a compound is found by adding together the molar masses of its elements. You can find the molar masses of each element on the Periodic Table provided in your Reference Tables.

Algebraically, the molar mass for calcium chloride, $CaCl_2$, can be expressed as:

$$[1 \times (\text{molar mass of calcium})] + [2 \times (\text{molar mass of chlorine})]$$

Plug in the molar masses of each element and solve:

$$(1 \times 40.1) + (2 \times 35.5) = \textbf{111.1 g/mol}$$

(NOTE: All values from **110. g/mol to 111.1 g/mol** were accepted as correct answers to this problem.)

57. An empirical formula is the formula of the actual compound reduced to the smallest whole number ratio. For example, the empirical formula of C_2H_4 is CH_2. To find out what the true formula is for the liquid compound, simply divide the gram-formula mass of the empirical formula into the true gram-formula (molar) mass of the compound to find the "scale-up factor."

The gram-formula mass of the empirical formula = 12.0 + 1.0 + 35.5 = 48.5 g/mol for the elements C, H, and Cl, respectively.

Divide this empirical formula mass into the true molar mass of the compound:

$$\frac{97}{48.5} = 2$$

This means that the molecular formula is twice as big as the empirical formula, in other words, $\textbf{C}_2\textbf{H}_2\textbf{Cl}_2$.

58. Since this question specifically asks you to answer "in terms of electrons," the correct answer will state how electrons are involved in the formation of an ionic bond. Ionic bonds form when **electrons are transferred from the metal (calcium) to the nonmetal (chlorine)**. When metals bond with nonmetals, electrons are completely transferred because of the large electronegativity difference between the two elements.

59. Unsaturated hydrocarbons contain at least one double or triple bond between carbon atoms. The term "unsaturated" implies that other atoms could be added to the compound by breaking the multiple bond between the carbon atoms. **The hydrocarbon is unsaturated because it has a double bond between the carbon atoms.**

60. Reference Table R is extremely helpful when classifying organic compounds. The product in the reaction shown is classified as a **halide (or halocarbon)** because of the presence of a halogen (Cl) on the molecule.

61. The metalloids are shown on the diagram of the Periodic Table below. These six metalloids must be committed to memory. **Triad 1** contains the metalloid tellurium (Te).

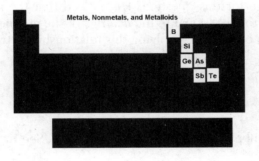

62. The elements in triad 2 are in Group 2 of the Periodic Table. Group 2 elements **each have two valence electrons**, which causes them to have the same chemical properties.

63. Consult Reference Table S for the first ionization energy data for most elements. The elements in triad 3 are all halogen elements in Group 17 of the Periodic Table. They are listed in order of increasing atomic number: Cl (17), Br (35), and I (53). Values for the first ionization energy for these elements are: 1251 kJ/mol, 1140 kJ/mol, and 1008 kJ/mol, respectively. **The first ionization energy decreases as the atomic number increases.** This is because as the atomic number increases, the number of electrons also increases, and the outermost (valence) electrons are located in electron shells that are farther from the nucleus. Less energy is required to remove electrons that are farther from the nucleus.

64. We are comparing equal masses of lithium and potassium, which are the first and third elements in triad 4. Density information from Reference Table S is needed to figure out how the 100. gram samples of these two elements will differ in volume. The element with the higher density will be capable of packing more mass in a smaller volume since the formula for density is $d = \dfrac{m}{V}$ (see Reference Table T). Lithium has a density of 0.534 g/cm^3, while potassium has a density of 0.89 g/cm^3. **Since potassium has a higher density, it can pack the same mass into a smaller volume.**

(NOTE: Alternatively, you can solve for the volume occupied by 100. g of each element using the density formula. **You will find that 100. g of Li occupies 187 cm^3, while 100. g of K occupies 112 cm^3.**)

65. Dobereiner's mathematical relationship for his element triads is stated in the reading passage for this question. He noticed that the atomic mass of the middle element is roughly equal to the *average* of the atomic masses of the first and last element in each triad. Applying this relationship to triad 2 yields the following mathematical expression:

$$\frac{40.\ u + 137\ u}{2}$$

(NOTE: The symbol u stands for atomic mass unit. See Reference Table D.)

PART C

[All questions in Part C are worth 1 point.]

66. An exothermic chemical reaction is a reaction that releases heat to the environment (heat "<u>ex</u>its" the system). Since **heat is on the product side of the equation**, the reaction must be exothermic.

67. Chemical changes take place when chemical bonds break and new bonds form, creating different substances than those that were there originally. Since **the products are different substances than the reactants**, a chemical change has taken place.

68. To find the percent composition by mass of carbon in $C_6H_{10}O_5$, use the percent composition formula from Reference Table T:

$$\% \text{ composition by mass} = \frac{\text{mass of part}}{\text{mass of whole}} \times 100$$

In this case, the "part" is the mass of carbon in the compound, and the "whole" is the gram-formula mass of the entire compound.

$$\% \text{ composition of carbon by mass} = \frac{6(12.0 \text{ g/mol})}{162.1 \text{ g/mol})} \times 100$$

(NOTE: It is not necessary to solve for the percent composition by mass of carbon in this problem since you are only asked for the numerical setup.)

69. To balance a chemical equation, the number of each type of atom must be equal on both sides of the equation. The unbalanced equation is shown below:

_____$CH_4(g)$ + _____ $H_2O(g)$ + energy → _____$CO(g)$ + _____$H_2(g)$

As written, there is 1 carbon on each side of the equation, so carbon is already balanced. There is also 1 oxygen on each side of the equation, so oxygen is already balanced. However, there are 6 hydrogen atoms on the left side of the reaction arrow and only 2 hydrogen atoms shown on the right side of the equation, so a coefficient of 3 is needed in front of the hydrogen on the right side of the equation to have 6 H atoms on both sides:

_____$\mathbf{CH_4(g)}$ + _____ $\mathbf{H_2O(g)}$ + **energy** → _____$\mathbf{CO(g)}$ + __3__$\mathbf{H_2(g)}$

70. When the temperature of this reaction increases, the particles move more quickly. This is consistent with the kinetic molecular theory of gases, which states that temperature is directly related to the average kinetic energy of gas particles. Faster moving particles will strike each other more often and with more force. Collision theory states that reactant particles must collide with enough force in order to react. By increasing the temperature, **more methane and steam molecules will have sufficient kinetic energy to react**.

71. Interval *A* represents **the heat of reaction**, or **ΔH**, which is defined as the potential energy of the products (on the right-hand side of the energy "hill") minus the potential energy of the reactants (on the left-hand side of the energy "hill"). Since the products are at a lower potential energy than the reactants, the reaction shown has a $-\Delta H$ and is therefore exothermic.

72. All coefficients in chemical reactions represent particles or moles. Mole ratios can be used to find the number of moles of any substance in a chemical equation if the moles of another substance are known. You could set up this problem as follows:

$$\frac{50.0 \; \cancel{\text{mol } N_2}}{1} \times \frac{3 \text{ mol } H_2}{1 \, \cancel{\text{mol } N_2}} = \textbf{150.0 mol } \textbf{H}_2$$

You could also set up mole ratios to solve this problem:

$$\frac{50.0 \text{ mol } N_2}{1} = \frac{x \text{ mol } H_2}{3}$$
$$x = \textbf{150.0 mol } \textbf{H}_2$$

73. Organic compounds always contain the element **carbon**.

74. There is a double covalent bond present in the reactant molecules. Since single covalent bonds consist of one shared pair of electrons, there are **2 pairs** of electrons in a double covalent bond (in other words, **4 electrons are shared**).

75. Synthesis reactions follow the general pattern A + B → AB. **Two reactants combine to form one product.**

76. Diethyl ether boils at 34.6°C, whereas water boils at 100°C. Diethyl ether does not require as much energy to separate its molecules because **diethyl ether has weaker intermolecular forces than water**.

77. Isomers are molecules with the same chemical formula but different structures. Since the problem asks for the same functional group with a different structure, you will need to draw an ether but change the position of the carbon atoms relative to the oxygen. Two possible isomers with the ether functional group are shown below:

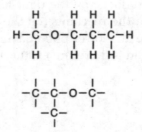

78. The presence of the battery is a sure sign that this is an electrolytic cell. **Electrolysis uses electrical energy to force a nonspontaneous chemical change to occur.**

79. All electrolytes dissociate into ions in aqueous solution. **The freely moving ions in this solution allow the conduction of electricity.**

80. The aqueous product in the reaction shown is NaOH (aq). Remember that the sum of the oxidation numbers in any compound must equal zero. The oxidation number of both Na and H in this compound is +1. These oxidation numbers can be found on the Periodic Table in your Reference Tables. The oxidation number of oxygen must be **−2** for the sum of the oxidation numbers to equal zero.

81. The pH value of the brine solution before the battery is connected is 7.0, as stated in the information given in the introductory paragraph for this problem. As the battery operates, the concentration of NaOH increases in the solution, and NaOH is a strong base (see Reference Table L). Since bases have pH values greater than 7, **the pH value will increase as this cell operates**.

82. The notation "Pb-214" is one way of describing a specific isotope of lead. This particular isotope has a mass number of 214, which means that the number of protons plus the number of neutrons is equal to 214. We can represent this mathematical relationship in the expression below:

$$\text{\# protons} + \text{\# neutrons} = 214$$

Since lead's atomic number is 82, all lead atoms have 82 protons. Substitute this value into the previous expression to solve for the number of neutrons:

$$82 + \text{\# neutrons} = 214$$

Rearrange the values:

$$214 - 82 = \text{\# neutrons}$$

There are **132 neutrons** in Pb-214.

83. To balance a nuclear equation, the sum of the atomic numbers (bottom) and the mass numbers (top) must be equal on both sides of the arrow.

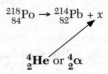

$$^{218}_{84}\text{Po} \rightarrow {}^{214}_{82}\text{Pb} + x$$

$${}^{4}_{2}\text{He or } {}^{4}_{2}\alpha$$

A mass number of 4 and an atomic number of 2 will correctly balance this nuclear equation. This is an alpha decay equation, and the particle that is emitted is an alpha particle.

84. From Reference Table *N*, the half-life of Rn-222 is 3.823 days. In 7.646 days, Rn-222 goes through two half-lives (7.646 days/3.823 days). One half-life is the time required for a radioisotope to decay to half of its original mass. After one half-life, only one half of the original Rn-222 will remain. After another half-life passes, only $\frac{1}{2}$ of this $\frac{1}{2}$ will remain. $\frac{1}{2} \times \frac{1}{2} = \frac{1}{4}$. The process of radioactive decay is represented below:

| Initial amount 100% | After 1st half-life, 50% remains | After 2nd half-life, 25% remains |

85. Look at the Periodic Table in your Reference Tables for the element bismuth (Bi). Bismuth has an atomic number of 83. Using this number and the mass number of 210, as given in the problem, the Decay Series graph shows that Bi-210 decays into polonium, which has an atomic number of 84. Since **a new element forms**, the decay of Bi-210 is considered a transmutation.

Mark (✔) the questions you answered correctly. Count the number of checks and follow the formulas given to determine your score on each topic.

Core Area   □ Questions Answered Correctly

50: (1)

Section M—Math Skills
□ Number of checks ÷ 1 × 100 = ___ %

55, 65: (2)

Section R—Reading Skills
□ Number of checks ÷ 2 × 100 = ___ %

1, 2, 3, 31, 32: (5)

Section I—Atomic Concepts
□ Number of checks ÷ 5 × 100 = ___ %

5, 6, 33, 61–64, 82: (8)

Section II—Periodic Table
□ Number of checks ÷ 8 × 100 = ___ %

34, 35, 56, 57, 68, 69, 72, 75: (8)

Section III—Moles/Stoichiometry
□ Number of checks ÷ 8 × 100 = ___ %

7–9, 11, 12, 48, 49, 51, 52, 58, 74, 76: (12)

Section IV—Chemical Bonding
□ Number of checks ÷ 12 × 100 = ___ %

4, 10, 13–19, 21–23, 36–41, 66, 67: (20)

Section V—Physical Behavior of Matter
□ Number of checks ÷ 20 × 100 = ___ %

20, 24, 53, 54, 70, 71: (6)

Section VI—Kinetics and Equilibrium
□ Number of checks ÷ 6 × 100 = ___ %

42, 43, 59, 60, 73, 77: (6)

Section VII—Organic Chemistry
□ Number of checks ÷ 6 × 100 = ___ %

Core Area	☐ Questions Answered Correctly

25, 26, 44, 78, 80: (5)

Section VIII—Oxidation–Reduction
☐ Number of checks ÷ 5 × 100 = ___ %

27, 45–47, 79, 81: (6)

Section IX—Acids, Bases, and Salts
☐ Number of checks ÷ 6 × 100 = ___ %

28–30, 83–85: (6)

Section X—Nuclear Chemistry
☐ Number of checks ÷ 6 × 100 = ___ %

Examination
June 2018
Chemistry—Physical Setting

PART A

Answer all questions in this part.

Directions (1–30): For *each* statement or question, write in the answer space the *number* of the word or expression that, of those given, best completes the statement or answers the question. Some questions may require the use of the *2011 Edition Reference Tables for Physical Setting/Chemistry*.

1 Which statement describes the charge and location of an electron in an atom?

 (1) An electron has a positive charge and is located outside the nucleus.

 (2) An electron has a positive charge and is located in the nucleus.

 (3) An electron has a negative charge and is located outside the nucleus.

 (4) An electron has a negative charge and is located in the nucleus. 1 _____

Table O

2 Which statement explains why a xenon atom is electrically neutral?

 (1) The atom has fewer neutrons than electrons.

 (2) The atom has more protons than electrons.

 (3) The atom has the same number of neutrons and electrons.

 (4) The atom has the same number of protons and electrons. 2 _____

Protons and electrons cancel out

3 If two atoms are isotopes of the same element, the atoms must have

   (1) the same number of protons and the same number of neutrons

   (2) the same number of protons and a different number of neutrons

   (3) a different number of protons and the same number of neutrons

   (4) a different number of protons and a different number of neutrons

3 _____

two isotopes cannot have the same # of neutrons

4 Which electrons in a calcium atom in the ground state have the greatest effect on the chemical properties of calcium?

   (1) the two electrons in the first shell

   (2) the two electrons in the fourth shell

   (3) the eight electrons in the second shell

   (4) the eight electrons in the third shell

4 _____

Valence effects properties the most

5 The weighted average of the atomic masses of the naturally occurring isotopes of an element is the

   (1) atomic mass of the element

   (2) atomic number of the element

   (3) mass number of each isotope

   (4) formula mass of each isotope

5 _____

Atomic mass is never an exact number!

6 Which element is classified as a metalloid?

   (1) Cr   (3) Sc

   (2) Cs   (4) Si

6 _____

Metaloid staircase on reference tables

7 Which statement describes a chemical property of iron?

   (1) Iron oxidizes.

   (2) Iron is a solid at STP.

   (3) Iron melts.

   (4) Iron is attracted to a magnet.

7 _____

Oxidation is a chemical change

8 Graphite and diamond are two forms of the same element in the solid phase that differ in their

   (1) atomic numbers   (3) electronegativities

   (2) crystal structures   (4) empirical formulas

8 _____

Tight packed carbon structure gives diamond its strength

9 Which ion has the largest radius?

(1) Br^- (3) F^-

(2) Cl^- (4) I^- 9 _____

10 Carbon monoxide and carbon dioxide have

(1) the same chemical properties and the same physical properties
(2) the same chemical properties and different physical properties
(3) different chemical properties and the same physical properties
(4) different chemical properties and different physical properties   10 _____

Chemical formula determines property

11 Based on Table S, which group on the Periodic Table has the element with the highest electronegativity?

(1) Group 1   (3) Group 17

(2) Group 2   (4) Group 18   11 _____

Flourine has the highest electronegativity

12 What is represented by the chemical formula $PbCl_2(s)$?

(1) a substance
(2) a solution
(3) a homogeneous mixture
(4) a heterogeneous mixture   12 _____

Substances are chemically combined

13 What is the vapor pressure of propanone at 50.°C?

(1) 37 kPa   (3) 83 kPa

(2) 50. kPa   (4) 101 kPa   13 _____

Reference table H

14 Which statement describes the charge distribution and the polarity of a CH_4 molecule?

(1) The charge distribution is symmetrical and the molecule is nonpolar.
(2) The charge distribution is asymmetrical and the molecule is nonpolar.
(3) The charge distribution is symmetrical and the molecule is polar.
(4) The charge distribution is asymmetrical and the molecule is polar.   14 _____

CH_4 is symmetrical and therefore non polar

15 In a laboratory investigation, a student separates colored compounds obtained from a mixture of crushed spinach leaves and water by using paper chromatography. The colored compounds separate because of differences in

(1) molecular polarity   (3) boiling point
(2) malleability   (4) electrical conductivity   15 _____

Stronger polarity climbs higher

16 Which phrase describes the motion and attractive forces of ideal gas particles?

(1) random straight-line motion and no attractive forces
(2) random straight-line motion and strong attractive forces
(3) random curved-line motion and no attractive forces
(4) random curved-line motion and strong attractive forces   16 _____

Ideal gases would not effect each other

17 At which temperature will $Hg(\ell)$ and $Hg(s)$ reach equilibrium in a closed system at 1.0 atmosphere?

(1) 234 K   (3) 373 K
(2) 273 K   (4) 630. K   17 _____

18 A molecule of any organic compound has at least one

(1) ionic bond   (3) oxygen atom
(2) double bond   (4) carbon atom   18 _____

All organics contain Carbon

19 A chemical reaction occurs when reactant particles

(1) are separated by great distances
(2) have no attractive forces between them
(3) collide with proper energy and proper orientation
(4) convert chemical energy into nuclear energy   19 _____

Collisions make reaction

20 Systems in nature tend to undergo changes toward

(1) lower energy and lower entropy
(2) lower energy and higher entropy
(3) higher energy and lower entropy
(4) higher energy and higher entropy   20 _____

Chaos is natural

21 Which formula can represent an alkyne?

(1) C_2H_4 (3) C_3H_4

(2) C_2H_6 (4) C_3H_6 21 _____

$C_n H_{2n-2}$

22 Given the formula representing a compound:

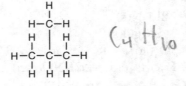

$C_4 H_{10}$

Which formula represents an isomer of this compound?

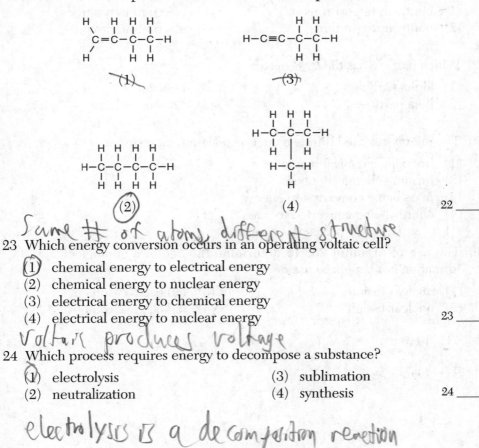

(1)   (3)

(2)   (4)   22 _____

Same # of atoms, different structure

23 Which energy conversion occurs in an operating voltaic cell?

(1) chemical energy to electrical energy

(2) chemical energy to nuclear energy

(3) electrical energy to chemical energy

(4) electrical energy to nuclear energy   23 _____

Voltaic produces voltage

24 Which process requires energy to decompose a substance?

(1) electrolysis   (3) sublimation

(2) neutralization   (4) synthesis   24 _____

electrolysis is a decomposition reaction

25 The concentration of which ion is increased when LiOH is dissolved in water?

(1) hydroxide ion   (3) hydronium ion
(2) hydrogen ion   (4) halide ion   25 _____

Name of the ion is on table E

26 Which equation represents neutralization?

(1) $6Li(s) + N_2(g) \rightarrow 2Li_3N(s)$
(2) $2Mg(s) + O_2(g) \rightarrow 2MgO(s)$
(3) $2KOH(aq) + H_2SO_4(aq) \rightarrow K_2SO_4(aq) + 2H_2O(\ell)$
(4) $Pb(NO_3)_2(aq) + K_2CrO_4(aq) \rightarrow 2KNO_3(aq) + PbCrO_4(s)$ 26 _____

acids and bases neutralize

27 The stability of an isotope is related to its ratio of

(1) neutrons to positrons   (3) electrons to positrons
(2) neutrons to protons   (4) electrons to protons   27 _____

28 Which particle has the *least* mass?

(1) alpha particle   (3) neutron
(2) beta particle   (4) proton   28 _____

29 The energy released during a nuclear reaction is a result of

(1) breaking chemical bonds
(2) forming chemical bonds
(3) mass being converted to energy
(4) energy being converted to mass   29 _____

Lost mass is converted into energy

30 The use of uranium-238 to determine the age of a geological formation is a beneficial use of

(1) nuclear fusion   (3) radioactive isomers
(2) nuclear fission   (4) radioactive isotopes   30 _____

Isotopes with long half lives can be used to date fossils

PART B–1

Answer all questions in this part.

Directions (31–50): For *each* statement or question, write in the answer space the *number* of the word or expression that, of those given, best completes the statement or answers the question. Some questions may require the use of the *2011 Edition Reference Tables for Physical Setting/Chemistry.*

Base your answers to questions 31 and 32 on your knowledge of chemistry and the bright-line spectra produced by four elements and the spectrum of a mixture of elements represented in the diagram below.

Bright-Line Spectra

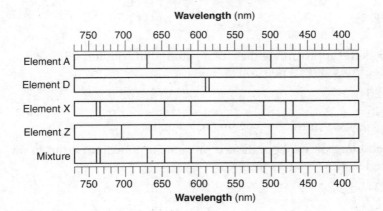

31 Which elements are present in this mixture?

(1) D and A   (3) X and A

(2) D and Z   (4) X and Z   31 _____

Match the lines

32 Each line in the spectra represents the energy

(1) absorbed as an atom loses an electron

(2) absorbed as an atom gains an electron

(3) released as an electron moves from a lower energy state to a higher energy state

(4) released as an electron moves from a higher energy state to a lower energy state   32 _____

Moving down releases energy

33 The table below shows the number of protons, neutrons, and electrons in four ions.

Four Ions

Ion	Number of Protons	Number of Neutrons	Number of Electrons
A	8	10	10
E	9	10	10
G	11	12	10
J	12	12	10

Which ion has a charge of 2−?

(1) A   (3) G
(2) E   (4) J   33 _____

– charges are determined by protons and electrons

34 What is the approximate mass of an atom that contains 26 protons, 26 electrons and 19 neutrons?

(1) 26 u   (3) 52 u
(2) 45 u   (4) 71 u   34 _____

Electrons don't effect mass

35 Which electron configuration represents a potassium atom in an excited state?

(1) 2-7-6   (3) 2-8-8-1
(2) 2-8-5   (4) 2-8-7-2   35 _____

If there is only one valence, another joins

36 What is the total number of neutrons in an atom of K-42?

(1) 19   (3) 23
(2) 20   (4) 42   36 _____

subtract atomic number from total mass

37 Given the equation representing a reaction:

$$2C + 3H_2 \rightarrow C_2H_6$$

What is the number of moles of C that must completely react to produce 2.0 moles of C_2H_6?

(1) 1.0 mol   (3) 3.0 mol
(2) 2.0 mol   (4) 4.0 mol   37 _____

moles on both sides must be equal

38 Given the equation representing a reaction:

$$Mg(s) + 2HCl(aq) \rightarrow MgCl_2(aq) + H_2(g)$$

Which type of chemical reaction is represented by the equation?

(1) synthesis (3) single replacement
(2) decomposition (4) double replacement   38 _____

Mg kicks out H₂

39 The table below lists properties of selected elements at room temperature.

Properties of Selected Elements at Room Temperature

Element	Density (g/cm³)	Malleability	Conductivity
sodium	0.97	yes	good
gold	19.3	yes	good
iodine	4.933	no	poor
tungsten	19.3	yes	good

Based on this table, which statement describes how two of these elements can be differentiated from each other?

(1) Gold can be differentiated from tungsten based on density.
(2) Gold can be differentiated from sodium based on malleability.
(3) Sodium can be differentiated from tungsten based on conductivity.
(4) Sodium can be differentiated from iodine based on malleability.   39 _____

Only choice without a common property

40 Which particle diagram represents a mixture?

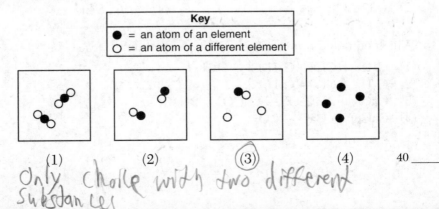

(1)   (2)   (3)   (4)   40 _____

Only choice with two different substances

41 An atom of which element reacts with an atom of hydrogen to form a bond with the greatest degree of polarity?

(1) carbon   (3) nitrogen
(2) fluorine   (4) oxygen   41 ____

Polarity is based on electronegativity difference

42 What is the concentration of an aqueous solution that contains 1.5 moles of NaCl in 500. milliliters of this solution?

(1) 0.30 M   (3) 3.0 M
(2) 0.75 M   (4) 7.5 M   42 ____

Moles of Solute / Liters of solution

43 The table below shows data for the temperature, pressure, and volume of four gas samples.

Data for Four Gases

Gas Sample	Temperature (K)	Pressure (atm)	Volume (L)
I	600.	2.0	5.0
II	300.	1.0	10.0
III	600.	3.0	5.0
IV	300.	1.0	10.0

Which two gas samples contain the same number of molecules?

(1) I and II   (3) II and III
(2) I and III   (4) II and IV   43 ____

44 Based on Table *I*, what is the ΔH value for the production of 1.00 mole of $NO_2(g)$ from its elements at 101.3 kPa and 298 K?

(1) +33.2 kJ   (3) +132.8 kJ
(2) −33.2 kJ   (4) −132.8 kJ   44 ____

Use reference tables

45 Which equation represents an addition reaction?

(1) $C_3H_8 + Cl_2 \rightarrow C_3H_7Cl + HCl$
(2) $C_3H_6 + Cl_2 \rightarrow C_3H_6Cl_2$
(3) $CaCl_2 + Na_2CO_3 \rightarrow CaCO_3 + 2NaCl$
(4) $CaCO_3 \rightarrow CaO + CO_2$ 45 ____

Addition adds a carbon bond

46 Given the balanced equation representing a reaction:

$$Ni(s) + 2HCl(aq) \rightarrow NiCl_2(aq) + H_2(g)$$

In this reaction, each Ni atom

(1) loses 1 electron   (3) gains 1 electron
(2) loses 2 electrons   (4) gains 2 electrons   46 _____

Count the difference

47 Which equation represents a reduction half-reaction?

(1) $Fe \rightarrow Fe^{3+} + 3e^-$ (3) $Fe^{3+} \rightarrow Fe + 3e^-$
(2) $Fe + 3e^- \rightarrow Fe^{3+}$ (4) $Fe^{3+} + 3e^- \rightarrow Fe$ 47 _____

reduction gains electrons

48 Given the balanced ionic equation representing a reaction:

$$Cu(s) + 2Ag^+(aq) \rightarrow Cu^{2+}(aq) + 2Ag(s)$$

During this reaction, electrons are transferred from

(1) $Cu(s)$ to $Ag^+(aq)$ (3) $Ag(s)$ to $Cu^{2+}(aq)$
(2) $Cu^{2+}(aq)$ to $Ag(s)$ (4) $Ag^+(aq)$ to $Cu(s)$ 48 _____

look at the change in charge

49 Which metal reacts spontaneously with Sr^{2+} ions?

(1) $Ca(s)$ (3) $Cs(s)$
(2) $Co(s)$ (4) $Cu(s)$ 49 _____

Reference table

50 Given the balanced equation representing a reaction:

$$HCl + H_2O \rightarrow H_3O^+ + Cl^-$$

The water molecule acts as a base because it

(1) donates an H^+ (3) donates an OH^-
(2) accepts an H^+ (4) accepts an OH^- 50 _____

Bases receive H+

PART B–2

Answer all questions in this part.

Directions (51–65): Record your answers on the answer sheet provided. Some questions may require the use of the *2011 Edition Reference Tables for Physical Setting/Chemistry*.

51 State the general trend in first ionization energy as the elements in Period 3 are considered from left to right.   [1]

52 Identify a type of strong intermolecular force that exists between water molecules, but does *not* exist between carbon dioxide molecules.   [1]

53 Draw a structural formula for 2-butanol.   [1]

Base your answers to questions 54 through 56 on the information below and on your knowledge of chemistry.

Some compounds of silver are listed with their chemical formulas in the table below.

Silver Compounds

Name	Chemical Formula
silver carbonate	Ag_2CO_3
silver chlorate	$AgClO_3$
silver chloride	$AgCl$
silver sulfate	Ag_2SO_4

54 Explain, in terms of element classification, why silver chloride is an ionic compound.   [1]

55 Show a numerical setup for calculating the percent composition by mass of silver in silver carbonate (gram-formula mass = 276 g/mol).   [1]

56 Identify the silver compound in the table that is most soluble in water.   [1]

Base your answers to questions 57 through 59 on the information below and on your knowledge of chemistry.

When a cobalt-59 atom is bombarded by a subatomic particle, a radioactive cobalt-60 atom is produced. After 21.084 years, 1.20 grams of an original sample of cobalt-60 produced remains unchanged.

57 Complete the nuclear equation by writing a notation for the missing particle. [1]

58 Based on Table N, identify the decay mode of cobalt-60. [1]

59 Determine the mass of the original sample of cobalt-60 produced. [1]

Base your answers to questions 60 through 62 on the information below and on your knowledge of chemistry.

A sample of a molecular substance starting as a gas at 206°C and 1 atm is allowed to cool for 16 minutes. This process is represented by the cooling curve below.

Cooling Curve for a Substance

60 Determine the number of minutes that the substance was in the liquid phase, only. [1]

61 Compare the strength of the intermolecular forces within this substance at 180.°C to the strength of the intermolecular forces within this substance at 120.°C.   [1]

62 Describe what happens to the potential energy and the average kinetic energy of the molecules in the sample during interval *DE*.   [1]

Base your answers to questions 63 through 65 on the information below and on your knowledge of chemistry.

The diagram below represents a cylinder with a moveable piston containing 16.0 g of $O_2(g)$. At 298 K and 0.500 atm, the $O_2(g)$ has a volume of 24.5 liters.

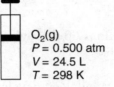

$O_2(g)$
$P = 0.500$ atm
$V = 24.5$ L
$T = 298$ K

63 Determine the number of moles of $O_2(g)$ in the cylinder. The gram-formula mass of $O_2(g)$ is 32.0 g/mol.   [1]

64 State the changes in *both* pressure and temperature of the gas in the cylinder that would increase the frequency of collisions between the $O_2(g)$ molecules.   [1]

65 Show a numerical setup for calculating the volume of $O_2(g)$ in the cylinder at 265 K and 1.00 atm.   [1]

PART C

Answer all questions in this part.

Directions (66–85): Record your answers on the answer sheet provided. Some questions may require the use of the *2011 Edition Reference Tables for Physical Setting/Chemistry*.

Base your answers to questions 66 through 69 on the information below and on your knowledge of chemistry.

In the late 1800s, Dmitri Mendeleev developed a periodic table of the elements known at that time. Based on the pattern in his periodic table, he was able to predict properties of some elements that had not yet been discovered. Information about two of these elements is shown in the table below.

Some Element Properties Predicted by Mendeleev

Predicted Elements	Property	Predicted Value	Actual Value
eka-aluminum (Ea)	density at STP	5.9 g/cm^3	5.91 g/cm^3
	melting point	low	30.°C
	oxide formula	Ea_2O_3	
	approximate molar mass	68 g/mol	
eka-silicon (Es)	density at STP	5.5 g/cm^3	5.3234 g/cm^3
	melting point	high	938°C
	oxide formula	EsO_2	
	approximate molar mass	72 g/mol	

66 Identify the phase of Ea at 310. K.   [1]

67 Write a chemical formula for the compound formed between Ea and Cl.   [1]

68 Identify the element that Mendeleev called eka-silicon, Es.   [1]

69 Show a numerical setup for calculating the percent error of Mendeleev's predicted density of Es.   [1]

Base your answers to questions 70 through 73 on the information below and your knowledge of chemistry.

Methanol can be manufactured by a reaction that is reversible. In the reaction, carbon monoxide gas and hydrogen gas react using a catalyst. The equation below represents this system at equilibrium.

$$CO(g) + 2H_2(g) \rightleftharpoons CH_3OH(g) + energy$$

70 State the class of organic compounds to which the product of the forward reaction belongs.   [1]

71 Compare the rate of the forward reaction to the rate of the reverse reaction in this equilibrium system.   [1]

72 Explain, in terms of collision theory, why increasing the concentration of $H_2(g)$ in this system will increase the concentration of $CH_3OH(g)$.   [1]

73 State the effect on the rates of both the forward and reverse reactions if no catalyst is used in the system.   [1]

Base your answers to questions 74 through 76 on the information below and on your knowledge of chemistry.

Fatty acids, a class of compounds found in living things, are organic acids with long hydrocarbon chains. Linoleic acid, an unsaturated fatty acid, is essential for human skin flexibility and smoothness. The formula below represents a molecule of linoleic acid.

$$
\begin{array}{c}
\text{H–C–C–C–C–C–C=C–C–C=C–C–C–C–C–C–C–C–C–O–H}
\end{array}
$$

74 Write the molecular formula of linoleic acid.   [1]

75 Identify the type of chemical bond between the oxygen atom and the hydrogen atom in the linoleic acid molecule.   [1]

76 On the diagram *on your answer sheet*, circle the organic acid functional group.   [1]

Base your answers to questions 77 through 79 on the information below and on your knowledge of chemistry.

Fuel cells are voltaic cells. In one type of fuel cell, oxygen gas, $O_2(g)$, reacts with hydrogen gas, $H_2(g)$, producing water vapor, $H_2O(g)$, and electrical energy. The unbalanced equation for this redox reaction is shown below.

$$H_2(g) + O_2(g) \rightarrow H_2O(g) + \text{energy}$$

A diagram of the fuel cell is shown below. During operation of the fuel cell, hydrogen gas is pumped into one compartment and oxygen gas is pumped into the other compartment. Each compartment has an inner wall that is a porous carbon electrode through which ions flow. Aqueous potassium hydroxide, $KOH(aq)$, and the porous electrodes serve as the salt bridge.

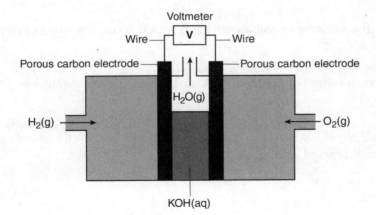

77 Balance the equation *on your answer sheet* for the reaction in this fuel cell, using the smallest whole-number coefficients. [1]

78 Determine the change in oxidation number for oxygen in this operating fuel cell. [1]

79 State the number of moles of electrons that are gained when 5.0 moles of electrons are lost in this reaction. [1]

Base your answers to questions 80 through 82 on the information below and on your knowledge of chemistry.

In a laboratory investigation, a student compares the concentration and pH value of each of four different solutions of hydrochloric acid, $HCl(aq)$, as shown in the table below.

Data for HCl(aq) Solutions

Solution	Concentration of HCl(aq) (M)	pH Value
W	1.0	0
X	0.10	1
Y	0.010	2
Z	0.0010	3

80 State the number of significant figures used to express the concentration of solution Z.   [1]

81 Determine the concentration of an $HCl(aq)$ solution that has a pH value of 4.   [1]

82 Determine the volume of 0.25 M $NaOH(aq)$ that would exactly neutralize 75.0 milliliters of solution X.   [1]

Base your answers to questions 83 through 85 on the information below and on your knowledge of chemistry.

Carbon dioxide is slightly soluble in seawater. As carbon dioxide levels in the atmosphere increase, more CO_2 dissolves in seawater, making the seawater more acidic because carbonic acid, $H_2CO_3(aq)$, is formed.

Seawater also contains aqueous calcium carbonate, $CaCO_3(aq)$, which is used by some marine organisms to make their hard exoskeletons. As the acidity of the seawater changes, the solubility of $CaCO_3$ also changes, as shown in the graph below.

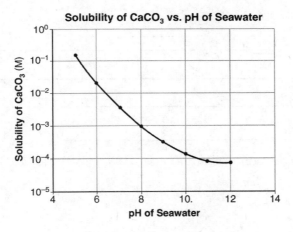

Solubility of CaCO$_3$ vs. pH of Seawater

83 State the trend in the solubility of $CaCO_3$ as seawater becomes more acidic.   [1]

84 State the color of bromcresol green in a sample of seawater in which the $CaCO_3$ solubility is 10^{-2} M.   [1]

85 A sample of seawater has a pH of 8. Determine the new pH of the sample if the hydrogen ion concentration is increased by a factor of 100.   [1]

Answer Sheet
June 2018
Chemistry—Physical Setting

PART B–2

51 Ionization energy increases

52 _____

53

54 Silver chloride is ionic because it
contains ionic bonds

7

55

56 $\underline{Ag\ ClO_3}$

57 $^{59}_{27}Co + \underline{^{1}_{0}n} \rightarrow ^{60}_{27}Co$

58 $\underline{\beta^-}$

59 $\underline{19.2}$ **g**

60 $\underline{5}$ **min**

61 _____

_____

_____

62 Potential energy: $\underline{Decreases}$

Average kinetic energy: $\underline{Remains\ the\ same}$

63 ___J___ **mol**

64 Change in pressure: _Increase_

Change in temperature: _Increase_

65

PART C

66 _____

67 _____

68 _____

69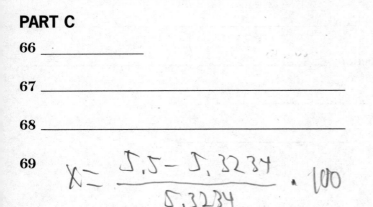

$$X = \frac{5.5 - 5.3234}{5.3234} \cdot 100$$

70 Alcohol

71 Rate of the forward reaction is equal to the rate of the reverse

72 If there is more H_2, it is more likely to collide and increase the concentration of CH_3OH

73 Rate of forward reaction: _Slower_

 Rate of reverse reaction: _Slower_

74 _____

75 _Polar covalent bond_

76

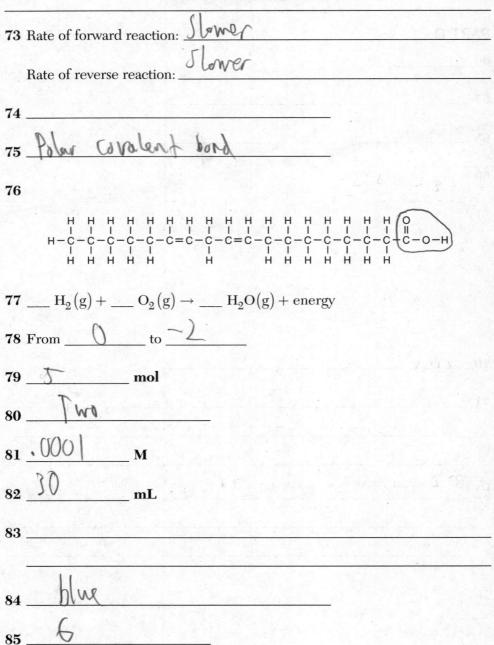

77 ___ $H_2(g)$ + ___ $O_2(g) \rightarrow$ ___ $H_2O(g)$ + energy

78 From _____0_____ to _____~2_____

79 _____5_____ **mol**

80 _____Two_____

81 _.0001_____ **M**

82 _30_____ **mL**

83 _____

 _____

84 _blue_

85 _6_

Answers
June 2018

Chemistry—Physical Setting

Answer Key

PART A

1. 3	7. 1	13. 3	19. 3	25. 1
2. 4	8. 2	14. 1	20. 2	26. 3
3. 2	9. 4	15. 1	21. 3	27. 2
4. 2	10. 4	16. 1	22. 2	28. 2
5. 1	11. 3	17. 1	23. 1	29. 3
6. 4	12. 1	18. 4	24. 1	30. 4

PART B–1

31. 3	35. 4	39. 4	43. 4	47. 4
32. 4	36. 3	40. 3	44. 1	48. 1
33. 1	37. 4	41. 2	45. 2	49. 3
34. 2	38. 3	42. 3	46. 2	50. 2

PART B–2 and **PART C**. *See* **Answers Explained**.

Answers Explained

PART A

1. **3** The basic structure of an atom is a positively charged nucleus surrounded by a negatively charged electron cloud, as shown in the illustration below. Reference Table *O* provides a summary of the names and charges of protons, neutrons, and electrons (which are identical to beta particles), but you do need to remember where each is located.

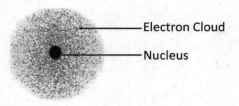

2. **4** Protons are positively charged, and electrons are negatively charged. All atoms have an equal number of protons and electrons, so atoms are electrically neutral (meaning that they have no charge at all).

3. **2** Isotopes are defined as two or more versions of the same element with different numbers of neutrons. For example, C-12 and C-14 are two isotopes of the element carbon. Since isotopes are the same element, they must have the same number of protons since the number of protons defines the element. However, isotopes have different mass numbers since the number of neutrons varies.

4. **2** According to the Periodic Table in your Reference Tables, calcium's electron configuration is 2-8-8-2. Calcium has two electrons in the fourth shell, which means that calcium has two valence electrons. Valence electrons are the outermost electrons in an atom. Valence electrons are involved in forming chemical bonds between atoms and are therefore largely responsible for an atom's chemical properties.

WRONG CHOICES EXPLAINED:
(1), (3), and (4) Calcium's electron configuration is 2-8-8-2. The electrons in the first, second, and third shells are referred to as "core" electrons that do not participate in forming bonds with other atoms.

5. **1** This is a "definition question." The atomic mass of an element is defined as the weighted average of the masses of its naturally occurring isotopes. Memorize this definition!

WRONG CHOICES EXPLAINED:

(2) The atomic number of an element equals the number of protons in the nucleus of any atom of this element.

(3) The mass number of an isotope is found by adding the number of protons to the number of neutrons in the nucleus of any atom of this isotope.

(4) Isotopes do not have formula masses since they consist of just one atom.

6. **4** Metalloids straddle the dividing line between metals and nonmetals. They exhibit properties of both metals and nonmetals. The six metalloid elements are boron (B), silicon (Si), arsenic (As), tellurium (Te), antimony (Sb), and germanium (Ge).

WRONG CHOICES EXPLAINED:

(1) Chromium (Cr) is a transition metal.

(2) Cesium (Cs) is an alkali metal (Group 1).

(3) Scandium (Sc) is a transition metal.

7. **1** Chemical properties describe how a substance will react with other substances. Oxidation is a type of chemical reaction in which a substance loses electrons.

WRONG CHOICES EXPLAINED:

(2), (3), and (4) These choices all describe physical properties that can be observed without changing the original substance (iron).

8. **2** Graphite and diamond are both allotropes of the element carbon. Allotropes are different structural forms of the same element. For example, graphite has a layered structure and diamonds have a three-dimensional network structure.

9. **4** The iodide ion, I^-, has the largest radius because it has the most electrons and therefore has the largest electron cloud.

WRONG CHOICES EXPLAINED:

(1), (2), and (3) The iodide ion contains electrons in the 5th electron shell whereas the other halogen ions mentioned in choices (1), (2), and (3) have no electrons in the 5th electron shell.

10. **4** Carbon monoxide is a linear, polar molecule whereas carbon dioxide is a linear, nonpolar molecule, as shown in the Lewis diagrams below:

:C≡O:

Carbon Monoxide

Ö=C=Ö

Carbon Dioxide

Since their molecular polarities are different, the two compounds will have different chemical and physical properties.

11. **3** When you consider the electronegativity of elements across any period of the Periodic Table, the element in Group 17 always has the highest value. These elements attract electrons more strongly than other elements in the same period. See this for yourself using the data provided on Reference Table *S*.

12. **1** All matter can be classified as either pure substances or mixtures. Pure substances can be either elements or compounds. $PbCl_2(s)$ is a compound.

WRONG CHOICES EXPLAINED:
(2) A solution is a homogeneous mixture, not a pure substance.
(3), (4) Homogeneous and heterogeneous mixtures are not pure substances.

13. **3** Use Reference Table *H* to read the vapor pressure of propanone at 50.°C.

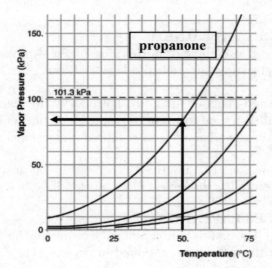

14. **1** The CH_4 molecule is a symmetrical shape known as a tetrahedron. Although each C−H bond is slightly polar, the bonds are symmetrically arranged around the central carbon atom, as shown in the diagram below.

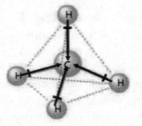

Since the four bonds are symmetrically arranged, electrons will be distributed symmetrically throughout the molecule and the overall molecule will be nonpolar.

15. **1** This question describes the laboratory separation technique known as paper chromatography, which may be used to separate components in some liquid mixtures. This process is illustrated in the diagram below.

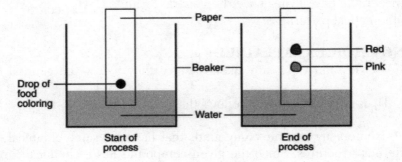

Molecules that have stronger polarity will be more attracted to polar solvents and will travel further up the paper than molecules that have weaker polarity.

16. **1** One of the principles of the kinetic molecular theory of gases is that gas particles move in random, constant, straight-line motion. They move in this way because of a second principle of the kinetic molecular theory of gases, namely that an ideal gas consists of particles that have no attractions to or repulsions from each other. Together, these two principles explain why gases expand to fill the space that they occupy.

17. **1** The solid and liquid phases of any pure substance can coexist at equilibrium only at a substance's melting point. According to Reference Table S, the element mercury (Hg) melts at 234 K.

18. **4** All organic compounds contain the element carbon. In fact, a simple definition of "organic chemistry" is the "chemistry of carbon."

19. **3** Collision theory states that there are two requirements for effective collisions (collisions between particles that result in a chemical reaction). These requirements are (1) that the colliding particles have sufficient energy (also known as the activation energy), and (2) that the colliding particles have proper orientation.

20. **2** Systems naturally tend to undergo changes that produce lower total energy and greater disorder or randomness (entropy). For example, when a glass topples off a table, it falls to the ground (a position of lower potential energy) and shatters (the particles in the glass have much greater randomness).

21. **3** Alkynes are one of three homologous series of hydrocarbons listed on Reference Table Q. Alkynes have the general formula C_nH_{2n-2}. C_3H_4 matches this mathematical pattern.

WRONG CHOICES EXPLAINED:
(1), (4) C_2H_4 and C_3H_6 are alkenes, since they follow the general formula C_nH_{2n}.
(2) C_2H_6 is an alkane, since it follows the general formula C_nH_{2n+2}.

22. **2** Isomers are organic compounds that have the same chemical formula but different structures. Since the given compound has a chemical formula of C_4H_{10}, its isomer must also have this formula.

WRONG CHOICES EXPLAINED:
(1) This compound's chemical formula is C_4H_8.
(3) This compound's chemical formula is C_4H_6.
(4) This compound's chemical formula is C_4H_{10}, but it is the exact same structure as the given compound and therefore is not an isomer. (Hint: If you can flip a structure over and match it exactly to another compound, the two compounds are exactly the same.)

23. **1** Voltaic cells, also known as batteries, use oxidation–reduction reactions to convert chemical energy to electrical energy.

24. **1** Decomposition is a type of chemical reaction. Energy can be applied to force decomposition to occur in a process known as electrolysis. For example, the electrolysis of water produces oxygen gas and hydrogen gas, as shown in the equation below:

$$2H_2O(\ell) + energy \rightarrow O_2(g) + 2H_2(g)$$

WRONG CHOICES EXPLAINED:
(2) Neutralization reactions happen when acids react with bases. These reactions are not decomposition reactions.
(3) Sublimation is a phase change directly from the solid phase to the gas phase. Phase changes are not chemical reactions.
(4) Synthesis reactions are the opposite of decomposition reactions.

25. **1** Lithium hydroxide, a base, dissociates in water to form lithium ions (Li^+) and hydroxide ions (OH^-). Remember, you can find the name of the OH^- ion on Reference Table E.

26. **3** A classic neutralization reaction follows the following general pattern:

$$Acid + Base \rightarrow Water + Salt \, (ionic \, compound)$$

KOH is a base (see Reference Table L), H_2SO_4 is an acid (see Reference Table K), and water and an ionic compound (K_2SO_4) are produced, so choice (3) is a neutralization reaction. None of the other answer choices contain acids and bases.

WRONG CHOICES EXPLAINED:
(1), (2) These are synthesis reactions.
(4) This is a double replacement reaction.

27. **2** The ratio of neutrons to protons in the nucleus of an atom determines whether the atom will be stable or not. If the ratio gets too large or too small, the nucleus will be unstable and will undergo radioactive decay.

28. **2** Information about all of the particles listed as answer choices for this question can be found on Reference Table O. The mass number is the top number on every symbol. Beta particles, which are identical to electrons, have a mass of 0 atomic mass units (amu).

(NOTE: The infinitesimal mass of an electron is much less than 1 amu, so it is considered insignificant.)

29. **3** Nuclear reactions are very different from chemical reactions. While the energy from chemical reactions comes from the formation of chemical bonds, the energy released from nuclear reactions comes from the conversion of a tiny bit of mass into energy according to the equation $E = mc^2$ (E is energy, m is mass, and c is the speed of light).

30. **4** Radioactive dating makes use of the consistent half-life of radio-isotopes to calculate the time elapsed during a certain event. According to Reference Table N, U-238 has a half-life of 4.47×10^9 years. By comparing the amount of U-238 remaining in a rock with the amount of its radioactive decay product, an approximate age for the rock can be determined.

WRONG CHOICES EXPLAINED:

(1), (2) Nuclear fusion and nuclear fission are important reactions for energy production.

(3) There is no such thing as a radioactive isomer! If you picked this answer choice, you were confused by the vocabulary. Consult the Glossary of Important Terms section of this book to relearn the difference between an "isomer" and an "isotope."

PART B–1

31. **3** All elements produce unique bright-line spectra when excited electrons in these atoms fall back to ground state. The bright-line spectrum of a mixture of two or more elements will contain all of the wavelength bands of each separate element superimposed together on the same spectrum. Element D can quickly be eliminated since the two wavelength bands at approximately 590 nm are not evident in the mixture. Element Z can also be eliminated since the wavelength band at approximately 450 nm is not found in the mixture. This leaves Elements X and A, and careful inspection shows that all wavelength bands from both elements are also found in the mixture.

32. **4** An atom in the excited state contains at least one electron that has previously absorbed energy and has moved to a higher energy orbital within the atom. When the electron returns to the lowest energy position possible, also known as the ground state, the extra energy that was absorbed is emitted (released) as a photon of electromagnetic energy. The diagram of all wavelengths of electromagnetic energy emitted from an atom is called its bright-line spectrum.

33. **1** Ions have charges because the number of electrons is either greater than or less than the number of protons. An imbalance of electrons and protons occurs when atoms lose or gain electrons during a chemical reaction. Electrons have a negative charge, and protons have a positive charge, so for an ion to have a charge of 2−, there would have to be 2 more electrons than protons.

WRONG CHOICES EXPLAINED:
(2) This ion has a charge of 1−, since the number of electrons is one greater than the number of protons.
(3) This ion has a charge of 1+, since the number of protons is one greater than the number of electrons.
(4) This ion has a charge of 2+, since the number of protons is two greater than the number of electrons.

34. **2** Protons and neutrons both have a mass of approximately 1 amu, while electrons are much tinier and their mass is ignored when calculating the mass of an atom. To find the mass of an atom (the mass number), add the number of protons to the number of neutrons:

$$26 \text{ protons} + 19 \text{ neutrons} = 45 \text{ u}$$

35. **4** The Periodic Table in your Reference Tables shows the ground-state configuration of potassium as 2-8-8-1. A potassium atom would be in an excited state if its configuration were 2-8-7-2 because this configuration shows that a third-shell electron has moved to the fourth shell.

WRONG CHOICES EXPLAINED:
(1), (2) There are only 15 electrons shown in these configurations. Potassium atoms always have 19 electrons. You might have chosen one of these answer choices if you forgot that the element symbol for potassium is K, not P (phosphorus).
(3) This is the ground-state configuration for a potassium atom. The question asked for an *excited*-state configuration.

36. **3** The mass number is defined as the sum of the protons and neutrons in an atom. Each proton and neutron has an approximate mass of 1 amu, so the sum of these particles will be equal to the mass expressed in amu. All potassium atoms have 19 protons (which equals the atomic number of potassium), and the isotope shown has a mass number of 42. The number of neutrons is found by subtracting the number of protons from the mass number:

$$42 - 19 = 23 \text{ neutrons}$$

37. **4** All coefficients in chemical reactions represent particles or moles. Mole ratios can be used to find the number of moles of any substance in a chemical equation if the moles of another substance are known. Two methods of setting up this problem are as follows:

$$\frac{2.0 \; \cancel{\text{mol } C_2H_6}}{1} \times \frac{2 \text{ mol C}}{1 \; \cancel{\text{mol } C_2H_6}} = 4.0 \text{ mol C}$$

OR

Set up mole ratios:

$$\frac{2.0 \text{ mol } C_2H_6}{1} = \frac{x \text{ mol C}}{2}$$
$$x = 4.0 \text{ mol C}$$

38. **3** Single replacement reactions follow the pattern $A + BC \rightarrow B + AC$. A more reactive element replaces a less reactive element in a compound. Single replacement reactions always have one uncombined element on each side of the equation.

39. **4** Elements can be differentiated, or distinguished, from one another on the basis of having different physical properties. In the data table provided, sodium is malleable whereas iodine is not. Malleability could be used to distinguish sodium from iodine.

40. **3** Mixtures are composed of two or more different substances that can be separated by physical means. Since they can be separated by physical means, the components of a mixture are not chemically bonded to one another.

WRONG CHOICES EXPLAINED:
(1), (2) These are diagrams of two different compounds. Compounds are pure substances, not mixtures.
(4) This is a diagram of an element. Elements are pure substances, not mixtures.

41. **2** The polarity of a bond depends upon the electronegativity difference between the elements in the bond. Electronegativity values for most elements are listed on Reference Table S. Fluorine has the highest electronegativity of all the elements, so an $H-F$ bond would have the greatest polarity.

42. **3** Molarity (M) is the concentration of a solution expressed as the number of moles of solute per liter of solution. The formula for calculating molarity is found on Reference Table T:

$$\text{molarity} = \frac{\text{moles of solute}}{\text{liter of solution}}$$

The given volume is in milliliters, and this needs to be converted into liters before plugging it into the molarity formula:

$$\frac{500. \cancel{mL}}{1} \times \frac{1\ L}{1,000\ \cancel{mL}} = 0.500\ L$$

Now, plug this volume into the molarity formula to solve for the molarity:

$$M_{NaCl} = \frac{1.5\ mol\ NaCl}{0.500\ L\ solution} = 3.0\ M$$

43. **4** Equal volumes of gas at the same conditions of temperature and pressure contain the same number of particles. This relationship is known as Avogadro's Law. For example, the molar volume of a gas at STP = 22.4 L/mol no matter what type of gas you have. Since gas samples II and IV have equal volumes and are at the same temperature and pressure, they will contain the same number of molecules.

44. **1** Reference Table *I* lists the heat of reaction (ΔH) for the production of *two* moles of $NO_2(g)$ as +66.4 kJ. For the production of just one mole of $NO_2(g)$, this value must be divided in two. The sign is still positive, however, because the reaction producing $NO_2(g)$ is endothermic no matter how many moles are produced.

45. **2** Addition reactions are organic reactions in which a diatomic element adds to a double or triple carbon–carbon bond. Addition reactions require an alkene or an alkyne as a reactant, and there is only one product since the multiple bond opens to accept the new atoms. Choice (2) is the only reaction that has an alkene or an alkyne as a reactant, and it is also the only reaction that produces just one product.

WRONG CHOICES EXPLAINED:
(1) This is a different organic reaction called a substitution reaction. A chlorine atom substitutes for a hydrogen atom on the organic compound. The organic reactant is an alkane, and there are two products.
(3) This is a double replacement reaction, which is not a type of organic reaction.
(4) This is a decomposition reaction, which is not a type of organic reaction.

46. **2** As a reactant, nickel is a pure element and has an oxidation number of 0. In the compound $NiCl_2$, each Ni atom has an oxidation number of 2+. Nickel atoms must lose two electrons to go from an oxidation number of 0 to an oxidation number of 2+.

47. **4** Reduction is defined as the gain of electrons. Half-reactions, like all chemical reactions, must be balanced for both mass and charge. The correct answer, $Fe^{3+} + 3e^- \rightarrow Fe$, shows iron(III) ions gaining electrons (they are reducing), and the sum of all charges on both sides of the arrow are equal (conservation of charge).

WRONG CHOICES EXPLAINED:
(1) This is an oxidation half-reaction.
(2) This shows iron atoms gaining electrons, which never happens. Metals are defined as elements that lose electrons when they bond. In addition, this half-reaction does not show conservation of charge (3 negative charges do not equal 3 positive charges).
(3) This is an incorrectly written oxidation half-equation that does not show conservation of charge (3 negative charges do not equal 3 positive charges).

48. **1** $Cu(s)$ loses two electrons to form Cu^{2+}. At the same time, each Ag^+ ion gains one electron to form $Ag(s)$. The transfer of electrons is from $Cu(s)$ to $Ag^+(aq)$.

WRONG CHOICES EXPLAINED:
(2), (3) These answer choices include the products of the reaction. The products are present when the reaction is over. The transfer of electrons is completed in order to form the products.

(4) If electrons were transferred from $Ag^+(aq)$ to $Cu(s)$, the silver ion would become more positive (which is impossible, since silver ions always have a $1+$ charge), and the $Cu(s)$ would become negative (which is impossible, since metals can only lose electrons when they bond).

49. **3** Reference Table J lists metals from the most active metals at the top to the least active metals at the bottom. Pure metal atoms never gain electrons, but metal ions can gain back the electrons they previously lost in the presence of a more active metal. Sr^{2+} ions will gain electrons back in the presence of $Cs(s)$, which is higher on Reference Table J and is therefore a more active metal.

WRONG CHOICES EXPLAINED:
(1), (2), and (4) These are all metals that are lower on Reference Table J than Sr. These metals are not as active as strontium and therefore would not be able to force electrons back onto the Sr^{2+} ion.

50. **2** Bases are proton (H^+) acceptors. Compare the reactant formula to the product formula. H_2O must accept a H^+ in order to form the hydronium ion H_3O^+.

PART B–2

[All questions in Part B–2 are worth 1 point.]

51. The **ionization energy increases** as the elements in Period 3 are considered from left to right across the Periodic Table. The increase in the ionization energy is caused by increasing nuclear charge. If you are struggling to remember this trend, you can answer this question using the data listed on Reference Table S.

52. Water molecules are attracted to each other by **hydrogen bonding**. Hydrogen bonding is an unusually strong intermolecular force that occurs when hydrogen is bonded to a small, highly electronegative atom such as F, O, or N. Molecules of carbon dioxide do not contain hydrogen and will be attracted by

weaker intermolecular forces. Water's strong hydrogen bonds explain why water is in the liquid state at room temperature, whereas carbon dioxide is a gas at room temperature.

53. Consult Reference Table *R* for help with questions about organic functional groups. The names of simple organic compounds contain all of the information necessary to figure out their structure. Start with the prefix of the compound 2-butanol. From Reference Table *P*, we see that 2-butanol contains 4 carbon atoms. Now look at the suffix. From Reference Table *R*, we find that compounds ending in the suffix −ol are alcohols and contain the functional group −OH. The number in front of the name of this compound gives the position of the functional group on the molecule. Putting all of this information together leads to the following structural formula for 2-butanol:

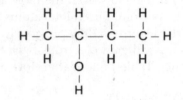

(NOTE: Other acceptable answers include the following:

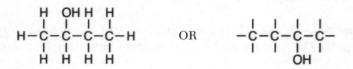

However, be careful to make sure to draw the bond on the second carbon specifically to the oxygen in the alcohol functional group. Students who drew imprecise bonds to the functional group did not receive credit.)

54. Silver chloride, AgCl, is classified as an ionic compound because bonds between metals and nonmetals are ionic bonds. **Silver is a metal, and chlorine is a nonmetal.**

(NOTE: <u>Be careful</u> on this type of question. Since you are asked to write your answer *in terms of element classification*, you must write about the kinds of elements included in this compound. Answering this question on the basis of the electronegativity difference between the elements would not be acceptable.)

55. Using the percent composition formula on Reference Table T:

$$\% \text{ composition by mass} = \frac{\text{mass of part}}{\text{mass of whole}} \times 100$$

$$\% \text{ composition by mass of silver in } Ag_2CO_3 = \frac{2(107.9 \text{ g/mol})}{276 \text{ g/mol}} \times 100$$

56. Reference Table F contains information about the solubility of various ionic compounds in water. Only the chlorate ion, ClO_3^-, is listed in the first column of this table, which lists ions that form *soluble* compounds in water. **$AgClO_3$** is therefore the most soluble compound listed in the table of Silver Compounds.

57. To balance a nuclear equation, the sum of the atomic numbers (bottom) and the mass numbers (top) must be equal on both sides of the arrow. Since the atomic numbers are the same on each side of the arrow, the bottom number on the missing particle must be 0. The top number on the missing particle must be 1, since $59 + 1 = 60$. Looking at Reference Table O, the missing particle must be a neutron $\left({}_{0}^{1}n\right)$.

$$ {}_{27}^{59}Co + \underline{\quad {}_{0}^{1}n \quad} \rightarrow {}_{27}^{60}CO $$

(NOTE: Many students missed this question because they did not notice that the arrow is *after* the particle. Be very careful to pay attention to the position of the arrow.)

58. The decay mode of Co-60, as listed on Reference Table N, is β^-, which is known as a **beta particle**. Remember to look at Reference Table O for the names and symbols of nuclear particles.
(NOTE: Other acceptable answers include ${}_{-1}^{0}e$, ${}_{-1}^{0}\beta$, or **beta decay**.)

59. From Reference Table N, the half-life of Co-60 is 5.271 years. In 21.084 years, Co-60 undergoes four half-lives (21.084 years/5.271 years $= 4$ half-lives). The information provided states that after 21.084 years, which is equivalent to four half-lives, only 1.20 grams of the original Co-60 sample remain. Remember, one half-life is the time required for a radioisotope to decay to half of its original mass. However, we need to think about this problem in the opposite way, since we need to work backward from the final mass remaining to calculate the original

mass. Working backward, before each half-life passes, there would be *twice as much* Co-60 present than after the half-life passes. The simplest approach to solving this problem is to take the 1.20 grams remaining after four half-lives pass and *double it four times,* as shown below:

1.20 g × 2 = 2.40 grams present before the 4th half-life
2.40 g × 2 = 4.80 grams present before the 3rd half-life
4.80 g × 2 = 9.6 grams present before the 2nd half-life
9.60 g × 2 = **19.20 grams** present originally, before the 1st half-life

60. The flat regions on the graph, intervals $\overline{BC}$ and $\overline{DE}$, represent times during the cooling period when the substance was changing phase. Since the information provided states that the substance begins as a gas, interval $\overline{BC}$ represents condensation, the phase change from a gas to a liquid. Therefore, during $\overline{CD}$, the substance was a liquid before it froze during $\overline{DE}$. Time period $\overline{CD}$ lasted **five minutes**.

61. Intermolecular forces are inversely proportional to the distance between particles. Particles that are closer together will be subject to far stronger intermolecular forces than particles that are far apart. A helpful analogy would be the strength of attraction between two magnets at different distances apart from each other. When the magnets are close together, they attract each other strongly, yet when they are moved far apart, they appear to have no attraction to each other whatsoever. Of course, the magnets *are* still attracted to each other, but at such a great distance, their attraction is imperceptible. Molecules behave similarly: **at 180°C as a gas, the molecules would experience weaker intermolecular forces than as a liquid at 120°C**, since the distance between the particles is much greater in the gas phase than in the liquid phase.

62. During interval $\overline{DE}$, the temperature does not change. Since the temperature measures the average kinetic energy of the particles in a sample of matter, if the temperature is not changing, then neither is the kinetic energy. However, during interval $\overline{DE}$, the particles in the substance are undergoing a phase change from the liquid state to the solid state. This phase change is exothermic, meaning that energy is released as the freezing process takes place. The energy released is potential energy: as the particles move closer together, the potential energy stored in the sample decreases. Therefore, **the average kinetic energy remains the same, and the potential energy decreases**.

63. Many students were thrown off when answering this question because of all of the information listed for this problem. The important fact is in the introductory information: the cylinder contains 16.0 grams of $O_2(g)$. This problem involves just a simple grams to moles conversion:

$$16.0 \; \cancel{g \; O_2(g)} \times \frac{1 \; mol \; O_2(g)}{32.0 \; \cancel{g \; O_2(g)}} = \textbf{0.500 mol } O_2(g)$$

64. One way to cause $O_2(g)$ molecules to collide more frequently is to push them closer together. $O_2(g)$ molecules would be forced to move closer together if the pressure on the cylinder were to increase. Another way to cause $O_2(g)$ molecules to collide more frequently is to increase their speed by raising the temperature of the $O_2(g)$ molecules. Therefore, the correct two changes are **increasing the pressure and increasing the temperature**.

65. Use the Combined Gas Law equation on Reference Table T to solve this problem:

$$\frac{P_1 V_1}{T_1} = \frac{P_2 V_2}{T_2}$$

Substitute information from the problem:

$$\frac{(\textbf{0.500 atm})(\textbf{24.5 L})}{\textbf{298 K}} = \frac{(\textbf{1.00 atm})(V_2)}{\textbf{265 K}}$$

You could also rearrange the equation as follows:

$$V_2 = \frac{(\textbf{0.500 atm})(\textbf{24.5 L})(\textbf{265 K})}{(\textbf{298 K})(\textbf{1.00 atm})}$$

PART C

[All questions in Part C are worth 1 point.]

66. Ea, the predicted element known as eka-aluminum, has a melting point of 30.°C. Convert this melting point to the Kelvin temperature scale using the formula found on Reference Table T:

$$K = °C + 273$$
$$\text{melting point of Ea} = 30°C + 273 = 303 \text{ K}$$

Since the given temperature in the problem is slightly greater than Ea's melting point, Ea would be a **liquid** at this temperature.

67. Ea makes an oxide that has the formula Ea_2O_3. If we assume that Ea is "like aluminum," then its oxide would be an ionic compound (a compound formed when a metal bonds with a nonmetal). In ionic compounds, the negative charges must exactly cancel out the positive charges since the overall compound is neutral in charge. This means that the Ea ion must have a charge of $+3$, since $2 \times +3$ will exactly cancel out oxygen's negative charge of 3×-2.
To determine the formula of the compound that is formed between Ea and Cl, first write charges on both ions:

$$Ea^{+3} \quad Cl^{-1}$$

Remember that ionic compounds must be neutral in overall charge. Three chloride ions are needed to cancel out the $+3$ charge of the Ea ion. Therefore, the chemical formula of the compound formed between Ea and Cl must be $\textbf{EaCl}_3$.
(NOTE: $GaCl_3$ was also accepted.)

68. Elements are often identified on the basis of physical properties. Reference Table S contains density and melting point data for many elements. First, use the equation given on Reference Table T to convert the given predicted melting point of $938°C$ into Kelvin so that it can be compared to the listed values on Reference Table S:

$$K = °C + 273$$
$$\text{melting point of Es} = 938°C + 273 = 1211 \text{ K}$$

Both the actual density of Es given in the problem and the actual melting temperature are equal to the values listed for the element **germanium (Ge)**.

69. The actual value for the density of Es (germanium) listed on the data table for this problem is 5.3234 g/cm^3. The predicted value was 5.5 g/cm^3. Use the percent error formula on Reference Table T to set up the calculation for percent error:

$$\% \text{ error} = \frac{\text{measured value} - \text{accepted value}}{\text{accepted value}} \times 100$$

Plug in the values from the problem:

$$\% \text{ error} = \frac{5.5 \text{ g/cm}^3 - 5.3234 \text{ g/cm}^3}{5.3234 \text{ g/cm}^3} \times 100$$

70. The product of the forward reaction, CH_3OH, is an **alcohol** because it contains the $-OH$ functional group. Reference Table R lists the organic functional groups.

71. Here is another "definition question." A chemical reaction is in a state of equilibrium when **the rate of the forward reaction equals the rate of the reverse reaction**. Memorize this definition!

72. The rate of a chemical reaction increases when the number of effective collisions between reactant particles increases. Effective collisions are collisions that have proper orientation and sufficient energy for a reaction to occur. Just by having more total collisions, there will also be more effective collisions since proper orientation will exist in more collisions. If the concentration of $H_2(g)$ increases, **there will be more effective collisions between CO(g) and H_2(g) molecules, producing more CH_3OH(g)**.

73. Do you remember what catalysts do? Adding a catalyst lowers the activation energy (E_a) for the reaction, making it possible for more collisions to have sufficient energy to react. Logically, removing a catalyst should *raise* the activation energy for *both* the forward reaction and the reverse reaction, making it *harder* for collisions to have sufficient energy to react. This should reduce the rate of reaction in both directions. Therefore, the **rate of the forward reaction will be slower** and the **rate of the reverse reaction will be slower**.

74. The molecular formula of a compound arranges the elements with subscripts, showing how many of each atom is in the compound (i.e., $C_xH_yO_z$). All that is required is to carefully count the number of each type of atom and write this information in the correct format. The answer to this question is $\mathbf{C_{18}H_{32}O_2}$.

(NOTE: Many students erroneously wrote a "condensed structural formula" to answer this question (i.e., $CH_3(CH_2)_4 (CH)_2...$, etc.). Knowing the names of the different styles for writing formulas is very important!)

75. Oxygen and hydrogen are two different nonmetals with different electronegativities. Oxygen has a much higher electronegativity than hydrogen, so it will attract the shared electrons in the bond between atoms much more strongly than hydrogen. Since the electrons are shared unequally, the bond between oxygen and hydrogen is **a polar covalent bond**.

76. The introductory information for this question states that the compound being considered, linoleic acid, is an organic acid. Reference Table R provides information about organic functional groups. Reference Table R shows that the general formula for the organic acid functional group is $-COOH$. Circle this part of the molecular diagram provided on the answer sheet, as shown below. Both diagrams are acceptable answers.

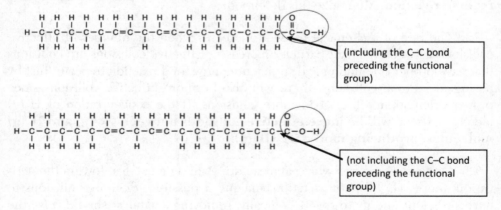

(including the C–C bond preceding the functional group)

(not including the C–C bond preceding the functional group)

(NOTE: The circled functional group *must* contain the hydrogen atom, since acids must contain hydrogen.)

77. To balance a chemical equation, the number of each type of atom must be equal on both sides of the equation. The unbalanced equation is shown below:

$$\underline{\quad} H_2(g) + \underline{\quad} O_2(g) \rightarrow \underline{\quad} H_2O(g) + energy$$

As shown above, there are 2 hydrogen atoms on each side of the equation, so hydrogen is already balanced. However, there are 2 oxygen atoms on the left of the reaction arrow and only 1 oxygen atom on the right side of the equation, so a coefficient of 2 is needed in front of water on the right side of the equation to have 2 O atoms on both sides:

$$\_\_ \; H_2(g) + \_\_ \; O_2(g) \rightarrow \underline{2} \; H_2O(g) + \text{energy}$$

With the coefficient of 2 in front of H_2O, hydrogen is no longer balanced. Adding a coefficient of 2 in front of the H_2 yields the final balanced equation:

$$\underline{2} \; \mathbf{H_2(g)} + \_\_ \; \mathbf{O_2(g)} \rightarrow \underline{2} \; \mathbf{H_2O(g)} + \textbf{energy}$$

78. The unbalanced redox reaction for the fuel cell, $H_2(g) + O_2(g) \rightarrow H_2O(g) + \text{energy}$, shows that as a reactant, oxygen is a pure and uncombined element. Pure and uncombined elements always have an oxidation number of 0. The reaction produces water, in which oxygen is covalently bonded to hydrogen. Remember that the sum of the oxidation numbers in any compound must equal zero. Oxidation numbers for elements can be found on the Periodic Table in your Reference Tables. The oxidation number of H in this compound is 1+. The oxidation number of oxygen must be 2− for the sum of the oxidation numbers to equal zero. Putting this all together, the oxidation number for oxygen changes from **0** to **2−**.

79. In a redox reaction, the number of electrons lost must equal the number of electrons gained. If 5.0 moles of electrons are lost, then **5.0 moles** of electrons must be gained.

80. The number 0.0010 has **two** significant figures. The zeros preceding the 1 in the thousandths place are referred to as "place-holding zeros" whose only purpose is to push the number over to the correct power of 10. The zero after the 1 in the thousandths place *is* significant, since it does not change the absolute value of the number but only serves to confirm that the number is precise to the ten-thousandths place. With decimals, it can be helpful to rewrite the number in scientific notation to make it easier to observe significant figures:

$$0.0010 = \underline{1.0} \times 10^{-3}$$

two significant digits

81. pH is a function of the base 10 exponent of the hydrogen ion concentration in an aqueous solution. The table provided for this question is very helpful in terms of recognizing the relationship between pH and HCl concentration. Notice that every time the concentration of HCl decreases by a factor of 10, the pH increases by 1. This is because pH is defined as the negative of the exponent of the base 10 concentration of the hydrogen ion. Using the pattern revealed in the data table, simply extend the table by one line to find that if a solution has a pH of 4, the concentration of HCl would be 10^{-4} M, which is **0.00010 M**.

82. Titration is the lab technique used to determine the concentration of an acid or a base by neutralizing it with an acid or a base of known concentration. The titration formula is found on Reference Table T:

$$M_A V_A = M_B V_B$$

in which M_A = molarity of H^+,
V_A = volume of acid,
M_B = molarity of OH^-,
and V_B = volume of base

Substitute the values from this problem:

$$(0.10 \text{ M})(75.0 \text{ mL}) = (0.25 \text{ M})(V_B)$$

HCl concentration in solution X

$$V_B = \frac{(0.10 \text{ M})(75.0 \text{ mL})}{0.25 \text{ M}} = \mathbf{30. \, mL}$$

83. The pH scale was developed to measure how acidic or basic a solution is, with pH values less than 7 indicating that a solution is acidic and pH values greater than 7 indicating that a solution is basic. As seawater becomes more acidic, the pH decreases. The graph shows that **as the pH decreases, the solubility of $CaCO_3$ increases**. Many students answered this question incorrectly because the graph shows descending values from left to right, but the decreasing solubility of $CaCO_3$ occurs as the pH increases (in other words, as it becomes more basic).

84. Use the graph to find that the pH of seawater is approximately 6.5 when the $CaCO_3$ solubility is 10^{-2} M. Reference Table M contains information about common acid–base indicators. The information about bromcresol green on this table states that the approximate pH for color change is 3.8–5.4 and the color

change is yellow to blue. This means that at pH values below 3.8, bromcresol green would be yellow. In the color change range of a pH of 3.8−5.4, the color would be changing from yellow to blue (in other words, green). Above a pH of 5.4, the color of bromcresol green would be **blue**.

85. pH measures the concentration of the hydrogen ion, H^+, in aqueous solution. pH is a logarithmic function, meaning that it is based on the exponent of the base 10 molarity of the hydrogen ion content. Each pH value is $10\times$ different in hydrogen ion molarity than the pH immediately higher and lower than itself. If the hydrogen ion concentration is increased by a factor of 100 ($= 10 \times 10$), the pH of the new solution will differ from the original value by 2. Hydrogen ions make a solution acidic, and since solutions that are more acidic have lower pH values, the new pH will be *lower* than the original. Therefore, $8 - 2 = $ **6** is the new pH.

Mark (✓) the questions you answered correctly. Count the number of checks and follow the formulas given to determine your score on each topic.

Core Area	☐ Questions Answered Correctly

69, 80: (2)

Section M—Math Skills
☐ Number of checks ÷ 2 × 100 = ___ %

28, 66, 68: (3)

Section R—Reading Skills
☐ Number of checks ÷ 3 × 100 = ___ %

1–5, 9, 31, 32, 34–36: (11)

Section I—Atomic Concepts
☐ Number of checks ÷ 11 × 100 = ___ %

6–8, 11, 39, 51: (6)

Section II—Periodic Table
☐ Number of checks ÷ 6 × 100 = ___ %

37, 38, 55, 63, 67, 74, 77: (7)

Section III—Moles/Stoichiometry
☐ Number of checks ÷ 7 × 100 = ___ %

14, 33, 41, 54, 56, 75: (6)

Section IV—Chemical Bonding
☐ Number of checks ÷ 6 × 100 = ___ %

10, 12, 13, 15–17, 40, 42, 43, 52, 60–62, 64, 65: (15)

Section V—Physical Behavior of Matter
☐ Number of checks ÷ 15 × 100 = ___ %

19, 20, 44, 71–73: (6)

Section VI—Kinetics and Equilibrium
☐ Number of checks ÷ 6 × 100 = ___ %

18, 21, 22, 45, 53, 70, 76: (7)

Section VII—Organic Chemistry
☐ Number of checks ÷ 7 × 100 = ___ %

Core Area	☐ Questions Answered Correctly

23, 24, 46–49, 78, 79: (8)

Section VIII—Oxidation–Reduction
☐ Number of checks $\div$ 8 $\times$ 100 = ___ %

25, 26, 50, 81–85: (8)

Section IX—Acids, Bases, and Salts
☐ Number of checks $\div$ 8 $\times$ 100 = ___ %

27, 29, 30, 57–59: (6)

Section X—Nuclear Chemistry
☐ Number of checks $\div$ 6 $\times$ 100 = ___ %

Examination August 2018
Chemistry—Physical Setting

PART A

Answer all questions in this part.

Directions (1–30): For *each* statement or question, write in the answer space the *number* of the word or expression that, of those given, best completes the statement or answers the question. Some questions may require the use of the *2011 Edition Reference Tables for Physical Setting/Chemistry*.

1 According to the wave-mechanical model, an orbital is defined as the most probable location of

   (1) a proton   (3) a positron

   (2) a neutron   (4) an electron   1 _____

2 The part of an atom that has an overall positive charge is called

   (1) an electron   (3) the first shell

   (2) the nucleus   (4) the valence shell   2 _____

3 Which subatomic particles each have a mass of approximately 1 u?

   (1) proton and electron   (3) neutron and electron

   (2) proton and neutron   (4) neutron and positron   3 _____

4 The discovery of the electron as a subatomic particle was a result of

   (1) collision theory

   (2) kinetic molecular theory

   (3) the gold-foil experiment

   (4) experiments with cathode ray tubes   4 _____

5 The elements on the Periodic Table of the Elements are arranged in order of increasing

(1) atomic mass   (3) atomic number

(2) formula mass   (4) oxidation number   5 _____

6 Which element is classified as a metalloid?

(1) Te   (3) Hg

(2) S   (4) I   6 _____

7 At STP, $O_2(g)$ and $O_3(g)$ are two forms of the same element that have

(1) the same molecular structure and the same properties

(2) the same molecular structure and different properties

(3) different molecular structures and the same properties

(4) different molecular structures and different properties   7 _____

8 Which substance can be broken down by chemical means?

(1) ammonia   (3) antimony

(2) aluminum   (4) argon   8 _____

9 Which statement describes $H_2O(\ell)$ and $H_2O_2(\ell)$?

(1) Both are compounds that have the same properties.

(2) Both are compounds that have different properties.

(3) Both are mixtures that have the same properties.

(4) Both are mixtures that have different properties.   9 _____

10 Which two terms represent major categories of compounds?

(1) ionic and nuclear

(2) ionic and molecular

(3) empirical and nuclear

(4) empirical and molecular   10 _____

11 Which formula represents an asymmetrical molecule?

(1) CH_4 (3) N_2

(2) CO_2 (4) NH_3 11 _____

12 Which statement describes the energy changes that occur as bonds are broken and formed during a chemical reaction?

(1) Energy is absorbed when bonds are both broken and formed.
(2) Energy is released when bonds are both broken and formed.
(3) Energy is absorbed when bonds are broken, and energy is released when bonds are formed.
(4) Energy is released when bonds are broken, and energy is absorbed when bonds are formed.   12 _____

13 A solid sample of copper is an excellent conductor of electric current. Which type of chemical bonds are in the sample?

(1) ionic bonds   (3) nonpolar covalent bonds
(2) metallic bonds   (4) polar covalent bonds   13 _____

14 Which list includes three forms of energy?

(1) thermal, nuclear, electronegativity
(2) thermal, chemical, electromagnetic
(3) temperature, nuclear, electromagnetic
(4) temperature, chemical, electronegativity   14 _____

15 Based on Table S, an atom of which element has the strongest attraction for electrons in a chemical bond?

(1) chlorine   (3) oxygen
(2) nitrogen   (4) selenium   15 _____

16 At which temperature and pressure would a sample of helium behave most like an ideal gas?

(1) 75 K and 500. kPa   (3) 300. K and 50. kPa
(2) 150. K and 500. kPa   (4) 600. K and 50. kPa   16 _____

17 A cube of iron at 20.°C is placed in contact with a cube of copper at 60.°C. Which statement describes the initial flow of heat between the cubes?

(1) Heat flows from the copper cube to the iron cube.
(2) Heat flows from the iron cube to the copper cube.
(3) Heat flows in both directions between the cubes.
(4) Heat does not flow between the cubes.   17 _____

18 Which sample at STP has the same number of atoms as 18 liters of Ne(g) at STP?

    (1) 18 moles of Ar(g)
    (2) 18 liters of Ar(g)
    (3) 18 grams of $H_2O(g)$
    (4) 18 milliliters of $H_2O(g)$ 18 _____

19 Compared to H_2S, the higher boiling point of H_2O is due to the

    (1) greater molecular size of water
    (2) stronger hydrogen bonding in water
    (3) higher molarity of water
    (4) larger gram-formula mass of water   19 _____

20 In terms of entropy and energy, systems in nature tend to undergo changes toward

    (1) lower entropy and lower energy
    (2) lower entropy and higher energy
    (3) higher entropy and lower energy
    (4) higher entropy and higher energy   20 _____

21 Amines, amides, and amino acids are categories of

    (1) isomers
    (2) isotopes
    (3) organic compounds
    (4) inorganic compounds   21 _____

22 A molecule of which compound has a multiple covalent bond?

    (1) CH_4
    (2) C_2H_4
    (3) C_3H_8
    (4) C_4H_{10} 22 _____

23 Which type of reaction produces soap?

    (1) polymerization
    (2) combustion
    (3) fermentation
    (4) saponification   23 _____

24 For a reaction system at equilibrium, LeChatelier's principle can be used to predict the

    (1) activation energy for the system
    (2) type of bonds in the reactants
    (3) effect of a stress on the system
    (4) polarity of the product molecules   24 _____

25 Which value changes when a Cu atom becomes a Cu^{2+} ion?

(1) mass number

(3) number of protons

(2) oxidation number

(4) number of neutrons

25 _____

26 Which reaction occurs at the anode in an electrochemical cell?

(1) oxidation

(3) combustion

(2) reduction

(4) substitution

26 _____

27 What evidence indicates that the nuclei of strontium-90 atoms are unstable?

(1) Strontium-90 electrons are in the excited state.

(2) Strontium-90 electrons are in the ground state.

(3) Strontium-90 atoms spontaneously absorb beta particles.

(4) Strontium-90 atoms spontaneously emit beta particles.

27 _____

28 Which nuclear emission is listed with its notation?

(1) gamma radiation, $_0^0\gamma$

(3) neutron, $_{-1}^0\beta$

(2) proton, $_2^4He$

(4) alpha particle, $_1^1H$

28 _____

29 The energy released by a nuclear fusion reaction is produced when

(1) energy is converted to mass

(2) mass is converted to energy

(3) heat is converted to temperature

(4) temperature is converted to heat

29 _____

30 Dating once-living organisms is an example of a beneficial use of

(1) redox reactions

(3) radioactive isotopes

(2) organic isomers

(4) neutralization reactions

30 _____

PART B–1

Answer all questions in this part.

Directions (31–50): For *each* statement or question, write in the answer space the *number* of the word or expression that, of those given, best completes the statement or answers the question. Some questions may require the use of the *2011 Edition Reference Tables for Physical Setting/Chemistry.*

31 What is the net charge of an ion that has 11 protons, 10 electrons, and 12 neutrons?

(1) 1+   (3) 1–

(2) 2+   (4) 2–

31 _____

32 Which electron configuration represents the electrons of an atom in an excited state?

(1) 2-5   (3) 2-5-1

(2) 2-8-5   (4) 2-6

32 _____

33 Which element is a liquid at 1000. K?

(1) Ag   (3) Ca

(2) Al   (4) Ni

33 _____

34 Which formula represents ammonium nitrate?

(1) NH_4NO_3 (3) $NH_4(NO_3)_2$

(2) NH_4NO_2 (4) $NH_4(NO_2)_2$

34 _____

35 The empirical formula for butene is

(1) CH_2 (3) C_4H_6

(2) C_2H_4 (4) C_4H_8

35 _____

36 Which equation represents a conservation of charge?

(1) $2Fe^{3+} + Al \rightarrow 2Fe^{2+} + Al^{3+}$

(2) $2Fe^{3+} + 2Al \rightarrow 3Fe^{2+} + 2Al^{3+}$

(3) $3Fe^{3+} + 2Al \rightarrow 2Fe^{2+} + 2Al^{3+}$

(4) $3Fe^{3+} + Al \rightarrow 3Fe^{2+} + Al^{3+}$

36 _____

37 Given the balanced equation representing a reaction:

$$2H_2 + O_2 \rightarrow 2H_2O + energy$$

Which type of reaction is represented by this equation?

(1) decomposition   (3) single replacement

(2) double replacement   (4) synthesis   37 _____

38 When a Mg^{2+} ion becomes a Mg atom, the radius increases because the Mg^{2+} ion

(1) gains 2 protons   (3) loses 2 protons

(2) gains 2 electrons   (4) loses 2 electrons   38 _____

39 The *least* polar bond is found in a molecule of

(1) HI   (3) HCl

(2) HF   (4) HBr   39 _____

40 A solution is prepared using 0.125 g of glucose, $C_6H_{12}O_6$, in enough water to make 250. g of total solution. The concentration of this solution, expressed in parts per million, is

(1) 5.00×10^1 ppm   (3) 5.00×10^3 ppm

(2) 5.00×10^2 ppm   (4) 5.00×10^4 ppm   40 _____

41 What is the amount of heat, in joules, required to increase the temperature of a 49.5-gram sample of water from 22°C to 66°C?

(1) 2.2×10^3 J   (3) 9.1×10^3 J

(2) 4.6×10^3 J   (4) 1.4×10^4 J   41 _____

42 A sample of a gas in a rigid cylinder with a movable piston has a volume of 11.2 liters at STP. What is the volume of this gas at 202.6 kPa and 300. K?

(1) 5.10 L   (3) 22.4 L

(2) 6.15 L   (4) 24.6 L   42 _____

43 The equation below represents a reaction between two molecules, X_2 and Z_2. These molecules form an "activated complex," which then forms molecules of the product.

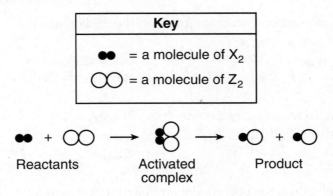

Which diagram represents the most likely orientation of X_2 and Z_2 when the molecules collide with proper energy, producing an activated complex?

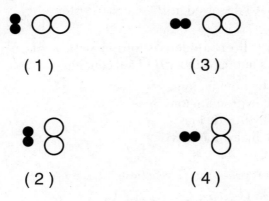

43 _____

44 What is the chemical name for the compound $CH_3CH_2CH_2CH_3$?

(1) butane   (3) decane
(2) butene   (4) decene

44 _____

45 In a laboratory activity, the density of a sample of vanadium is determined to be 6.9 g/cm³ at room temperature. What is the percent error for the determined value?

(1) 0.15%   (3) 13%
(2) 0.87%   (4) 15%

45 _____

46 Given the equation representing a reaction:

$$Cd + NiO_2 + 2H_2O \rightarrow Cd(OH)_2 + Ni(OH)_2$$

Which half-reaction equation represents the oxidation in the reaction?

(1) $Ni^{4+} + 2e^- \rightarrow Ni^{2+}$ (3) $Cd \rightarrow Cd^{2+} + 2e^-$

(2) $Ni^{4+} \rightarrow Ni^{2+} + 2e^-$ (4) $Cd + 2e^- \rightarrow Cd^{2+}$ 46 _____

47 Which metal reacts spontaneously with $NiCl_2(aq)$?

(1) $Au(s)$ (3) $Sn(s)$

(2) $Cu(s)$ (4) $Zn(s)$ 47 _____

48 Which solution is the best conductor of an electric current?

(1) 0.001 mole of NaCl dissolved in 1000. mL of water

(2) 0.005 mole of NaCl dissolved in 1000. mL of water

(3) 0.1 mole of NaCl dissolved in 1000. mL of water

(4) 0.05 mole of NaCl dissolved in 1000. mL of water   48 _____

49 Compared to a 1.0-liter aqueous solution with a pH of 7.0, a 1.0-liter aqueous solution with a pH of 5.0 contains

(1) 10 times more hydronium ions

(2) 100 times more hydronium ions

(3) 10 times more hydroxide ions

(4) 100 times more hydroxide ions   49 _____

50 Given the equation representing a reaction:

$$^{208}_{82}Pb + ^{70}_{30}Zn \rightarrow ^{277}_{112}Cn + ^{1}_{0}n$$

Which type of reaction is represented by this equation?

(1) neutralization   (3) substitution

(2) polymerization   (4) transmutation   50 _____

PART B–2

Answer all questions in this part.

Directions (51–65): Record your answers on the answer sheet provided. Some questions may require the use of the *2011 Edition Reference Tables for Physical Setting/Chemistry*.

51 Show a numerical setup for calculating the percent composition by mass of oxygen in Al_2O_3 (gram-formula mass = 102 g/mol).   [1]

52 Identify a laboratory process that can be used to separate a liquid mixture of methanol and water, based on the differences in their boiling points.   [1]

Base your answers to questions 53 through 55 on the information below and on your knowledge of chemistry.

The table below shows data for three isotopes of the same element.

Data for Three Isotopes of an Element

Isotopes	Number of Protons	Number of Neutrons	Atomic Mass (u)	Natural Abundance (%)
Atom D	12	12	23.99	78.99
Atom E	12	13	24.99	10.00
Atom G	12	14	25.98	11.01

53 Explain, in terms of subatomic particles, why these three isotopes represent the same element.   [1]

54 State the number of valence electrons in an atom of isotope *D* in the ground state.   [1]

55 Compare the energy of an electron in the first electron shell to the energy of an electron in the second electron shell in an atom of isotope *E*.   [1]

Base your answers to questions 56 through 58 on the information below and on your knowledge of chemistry.

The elements in Group 2 on the Periodic Table can be compared in terms of first ionization energy, electronegativity, and other general properties.

56 Describe the general trend in electronegativity as the metals in Group 2 on the Periodic Table are considered in order of increasing atomic number.   [1]

57 Explain, in terms of electron configuration, why the elements in Group 2 have similar chemical properties.   [1]

58 Explain, in terms of atomic structure, why barium has a lower first ionization energy than magnesium.   [1]

Base your answers to questions 59 through 61 on the information below and on your knowledge of chemistry.

A saturated solution of sulfur dioxide is prepared by dissolving $SO_2(g)$ in 100. g of water at 10.°C and standard pressure.

59 Determine the mass of SO_2 in this solution.   [1]

60 Based on Table G, state the general relationship between solubility and temperature of an aqueous SO_2 solution at standard pressure.   [1]

61 Describe what happens to the solubility of $SO_2(g)$ when the pressure is increased at constant temperature.   [1]

Base your answers to questions 62 through 65 on the information below and on your knowledge of chemistry.

Starting as a solid, a sample of a molecular substance is heated, until the entire sample of the substance is a gas. The graph below represents the relationship between the temperature of the sample and the elapsed time.

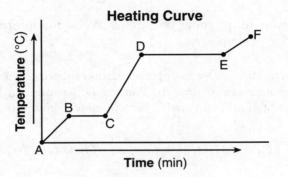

Heating Curve

62 Using the key *on your answer sheet*, draw a particle diagram to represent the sample during interval *AB*. Your response must include *at least six* molecules. [1]

63 Compare the average kinetic energy of the molecules of the sample during interval *BC* to the average kinetic energy of the molecules of the sample during interval *DE*. [1]

64 On the graph *on your answer sheet*, mark an **X** on the axis labeled "Temperature (°C)" to indicate the boiling point of the substance. [1]

65 State evidence that indicates the sample undergoes only physical changes during this heating. [1]

PART C

Answer all questions in this part.

Directions (66–85): Record your answers on the answer sheet provided. Some questions may require the use of the *2011 Edition Reference Tables for Physical Setting/Chemistry*.

Base your anwers to questions 66 through 68 on the information below and on your knowledge of chemistry.

"Water gas," a mixture of hydrogen and carbon monoxide, is an industrial fuel and source of commercial hydrogen. Water gas is produced by passing steam over hot carbon obtained from coal. The equation below represents this system at equilibrium:

$$C(s) + H_2O(g) + heat \rightleftharpoons CO(g) + H_2(g)$$

66 State, in terms of the rates of the forward and reverse reactions, what occurs when dynamic equilibrium is reached in this system. [1]

67 In the space *on your answer sheet*, draw a Lewis electron-dot diagram for a molecule of H_2O. [1]

68 Explain, in terms of collisions, why increasing the surface area of the hot carbon increases the rate of the forward reaction. [1]

Base your answers to questions 69 through 71 on the information below and on your knowledge of chemistry.

In a laboratory activity, each of four different masses of $KNO_3(s)$ is placed in a separate test tube that contains 10.0 grams of H_2O at 25°C.

When each sample is first placed in the water, the temperature of the mixture decreases. The mixture in each test tube is then stirred while it is heated in a hot water bath until all of the $KNO_3(s)$ is dissolved. The contents of each test tube are then cooled to the temperature at which KNO_3 crystals first reappear. The procedure is repeated until the recrystallization temperatures for each mixture are consistent, as shown in the table below.

Data Table for the Laboratory Activity

Mixture	Mass of KNO_3 (g)	Mass of H_2O (g)	Temperature of Recrystallization (°C)
1	4.0	10.0	24
2	5.0	10.0	32
3	7.5	10.0	45
4	10.0	10.0	58

69 Based on Table *I*, explain why there is a *decrease* in temperature when the $KNO_3(s)$ was first dissolved in the water.   [1]

70 Determine the percent by mass concentration of KNO_3 in mixture 2 after heating.   [1]

71 Compare the freezing point of mixture 4 at 1.0 atm to the freezing point of water at 1.0 atm.   [1]

Base your answers to questions 72 through 74 on the information below and on your knowledge of chemistry.

The balanced equation below represents the reaction between carbon monoxide and oxygen to produce carbon dioxide.

$$2CO(g) + O_2(g) \rightarrow 2CO_2(g) + \text{energy}$$

72 On the potential energy diagram *on your answer sheet*, draw a double-headed arrow ($\updownarrow$) to indicate the interval that represents the heat of reaction. [1]

73 Determine the number of moles of $O_2(g)$ needed to completely react with 8.0 moles of $CO(g)$. [1]

74 On the potential energy diagram *on your answer sheet*, draw a dashed line to show how the potential energy diagram changes when the reaction is catalyzed. [1]

Base your answers to questions 75 through 77 on the information below and on your knowledge of chemistry.

The equation below represents an industrial preparation of diethyl ether.

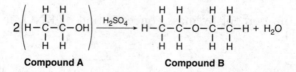

        Compound A   **Compound B**

75 Write the name of the class of organic compounds to which compound *A* belongs. [1]

76 Identify the element in compound *B* that makes it an organic compound. [1]

77 Explain, in terms of elements, why compound *B* is *not* a hydrocarbon. [1]

Base your answers to questions 78 through 81 on the information below and on your knowledge of chemistry.

A student is to determine the concentration of an NaOH(aq) solution by performing two different titrations. In a first titration, the student titrates 25.0 mL of 0.100 M H_2SO_4(aq) with NaOH(aq) of unknown concentration.

In a second titration, the student titrates 25.0 mL of 0.100 M HCl(aq) with a sample of the NaOH(aq). During this second titration, the volume of the NaOH(aq) added and the corresponding pH value of the reaction mixture is measured. The graph below represents the relationship between pH and the volume of the NaOH(aq) added for this second titration.

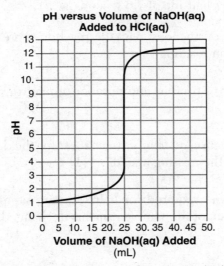

pH versus Volume of NaOH(aq) Added to HCl(aq)

78 Identify the positive ion present in the H_2SO_4(aq) solution before the titration. [1]

79 Complete the equation *on your answer sheet* for the neutralization that occurs in the first titration by writing a formula of the missing product. [1]

80 Based on the graph, determine the volume of NaOH(aq) used to exactly neutralize the HCl(aq). [1]

81 State the color of phenolphthalein indicator if it were added after the HCl(aq) was titrated with 50. mL of NaOH(aq). [1]

Base your answers to questions 82 through 85 on the information below and on your knowledge of chemistry.

When uranium-235 nuclei are bombarded with neutrons, many different combinations of smaller nuclei can be produced. The production of neodymium-150 and germanium-81 in one of these reactions is represented by the equation below.

$$\,_0^1\text{n} + \,_{92}^{235}\text{U} \rightarrow \,_{60}^{150}\text{Nd} + \,_{32}^{81}\text{Ge} + 5\,_0^1\text{n}$$

Germanium-81 and uranium-235 have different decay modes. Ge-81 emits beta particles and has a half-life of 7.6 seconds.

82 Explain, in terms of nuclides, why the reaction represented by the nuclear equation is a fission reaction.   [1]

83 State the number of protons and number of neutrons in a neodymium-150 atom.   [1]

84 Complete the equation *on your answer sheet* for the decay of Ge-81 by writing a notation for the missing nuclide.   [1]

85 Determine the time required for a 16.00-gram sample of Ge-81 to decay until only 1.00 gram of the sample remains unchanged.   [1]

Answer Sheet
August 2018
Chemistry—Physical Setting

PART B–2

51

52 _____

53 _____

_____

_____

54 _____

55 _____

_____

_____

56 _____

_____

_____

57 _____

_____

_____

58 _____

_____

_____

59 _____ g

60 _____

_____

_____

61 _____

_____

_____

62

Key
○ = a molecule of the substance

63 _____

_____

_____

64

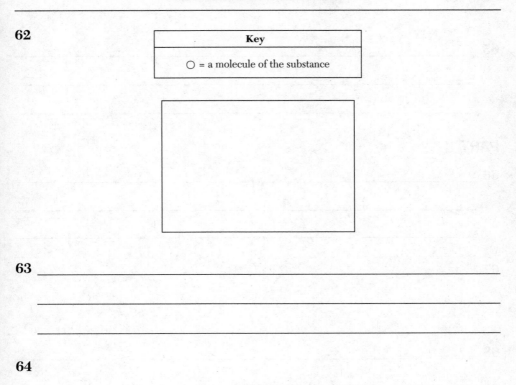

Heating Curve

65 _____

_____

_____

PART C

66 _____

_____

_____

67

68 _____

_____

_____

69 _____

_____

_____

70 _____ %

71 _____

_____

_____

72

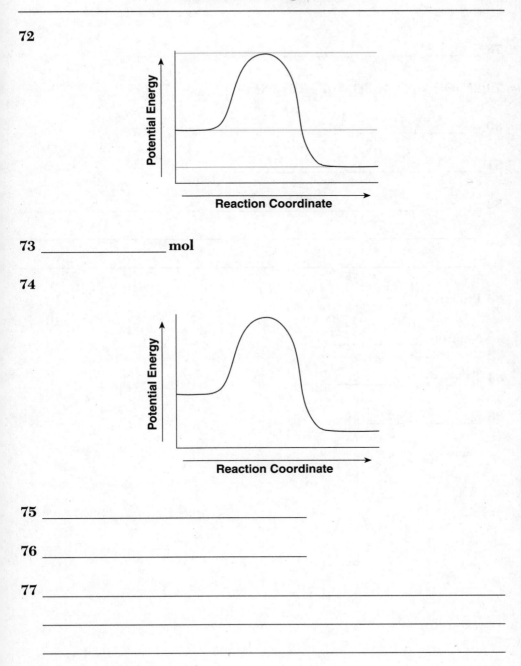

73 _____ **mol**

74

75 _____

76 _____

77 _____

_____

_____

78 _____

79 $2NaOH(aq) + H_2SO_4(aq) \rightarrow 2H_2O(\ell) +$ _____ (aq)

80 _____ **mL**

81 _____

82 _____

_____

_____

83 Protons: _____

 Neutrons: _____

84 $^{81}_{32}Ge \rightarrow \,^{0}_{-1}e +$ _____

85 _____ **s**

Answers
August 2018

Chemistry—Physical Setting

Answer Key

PART A

1. 4	7. 4	13. 2	19. 2	25. 2
2. 2	8. 1	14. 2	20. 3	26. 1
3. 2	9. 2	15. 3	21. 3	27. 4
4. 4	10. 2	16. 4	22. 2	28. 1
5. 3	11. 4	17. 1	23. 4	29. 2
6. 1	12. 3	18. 2	24. 3	30. 3

PART B–1

31. 1	35. 1	39. 1	43. 2	47. 4
32. 3	36. 4	40. 2	44. 1	48. 3
33. 2	37. 4	41. 3	45. 4	49. 2
34. 1	38. 2	42. 2	46. 3	50. 4

PART B–2 and **PART C**. *See* **Answers Explained**.

Answers Explained

PART A

1. **4** Electrons in an atom exist in the electron cloud surrounding the nucleus, where the protons and the neutrons are found. According to the wave-mechanical model of the atom, the electron cloud contains many different orbitals, which are three-dimensional regions of space where there is a high probability of finding electrons.

2. **2** Protons have a positive charge and are located in the nucleus of the atom. Refer to Reference Table O for help remembering the charges of subatomic particles.

WRONG CHOICES EXPLAINED:
(1) Electrons have a negative charge.
(3), (4) Electrons are found outside of the nucleus in shells or principal energy levels.

3. **2** Protons and neutrons both have a mass of approximately 1 atomic mass unit (u). Refer to Reference Table O to verify the masses of subatomic particles. The mass is the upper number associated with each symbol.

4. **4** J. J. Thomson discovered the electron during his famous cathode ray tube experiment. What first appeared to be light in the tube was deflected toward the positive side of an external electric field, proving that the light was caused by negatively charged particles.

5. **3** Each element has a unique atomic number that equals the number of protons in each atom. The Periodic Table is arranged from hydrogen (H, atomic number = 1) to ununoctium (Uuo, atomic number = 118) in order of increasing atomic number.

6. **1** The bolded staircase divider printed on the Periodic Table in your Reference Tables divides the metal elements from the nonmetal elements. Metalloids, elements with properties of both metals and nonmetals, are found next to this dividing line. Six elements are classified as metalloids: B, Si, As, Te, Ge, and Sb. Memorize these elements!

7. **4** O_2 and O_3 are both allotropes of the element oxygen. Allotropes are different physical forms of the same element. As seen in the structures that follow, O_2 is a linear, nonpolar molecule, while O_3 (ozone) is a bent, polar molecule. Their different structures cause these allotropes of oxygen to have different physical and chemical properties.

$$\ddot{O}=\ddot{O} \qquad \ddot{\ddot{O}}\diagdown \overset{\displaystyle \ddot{O}}{} \diagup \diagdown \ddot{O}$$

8. **1** Ammonia (NH_3) is a chemical compound that can be broken apart into its elements by a chemical reaction.

WRONG CHOICES EXPLAINED:
(2), (3), and (4) All of these answer choices are elements found on the Periodic Table. Elements are the simplest pure substances and cannot be broken down further by chemical means.

9. **2** H_2O is water, and H_2O_2 is hydrogen peroxide. These are two different compounds. The state symbol (ℓ) represents a liquid that is a pure compound, not a mixture. Since the molecules of these two compounds differ in size, composition, and shape, the intermolecular forces within these substances will be different, resulting in different physical properties.

WRONG CHOICES EXPLAINED:
(1) Different compounds will not have the same properties because their molecules are not the same.
(3), (4) The state symbol for mixtures containing water is (aq), meaning an aqueous solution.

10. **2** Compounds are distinguished by the type of bonding that holds them together. Ionic compounds contain particles held together by ionic bonds, and molecular compounds contain particles held together by covalent bonds.

11. **4** The best way to determine whether or not a molecule is symmetrical is to draw its Lewis structure and evaluate the bonding. The Lewis structure for ammonia is as follows:

$$H—\overset{\displaystyle ..}{N}—H$$
$$|$$
$$H$$

The lone pair of electrons on the central atom pushes away the bonding pairs of electrons, causing this molecule to have an asymmetrical, trigonal pyramidal shape:

WRONG CHOICES EXPLAINED:
   (1) CH_4 molecules are tetrahedral and symmetrical.
   (2) CO_2 molecules are linear and symmetrical.
   (3) N_2 molecules are linear and symmetrical.

12. **3** Energy must be *absorbed* to break the bonds that hold atoms together, just as energy would be needed to pull apart two magnets that are attracted to each other. However, when chemical bonds form, energy is *released* since atoms are more stable once they are bonded together.

13. **2** Copper is a metal element that is held together by metallic bonds. Metals are great conductors of electricity because of the mobile "sea of electrons" that is created when metallic bonds form.

14. **2** Energy can exist in many different forms. The motion of atoms and molecules within a substance creates thermal energy. Chemical energy is stored in chemical bonds, and electromagnetic energy is energy that travels through space in the form of waves.

WRONG CHOICES EXPLAINED:
   (1), (4) Electronegativity is not a form of energy. Electronegativity is the amount of attraction that an atom has for a bonded pair of electrons.
   (3), (4) Temperature is not a form of energy. Temperature is directly related to the kinetic energy of particles in a sample of matter.

15. **3** This question is asking about electronegativity, which is defined as an atom's attraction for the electrons in a chemical bond. Look up electronegativity values for these elements in Reference Table *S*. Oxygen has an electronegativity of 3.4, the second highest value on the table!

16. **4** Real gases behave more like ideal gases when the temperature is high and the pressure is low. These conditions minimize any possible attractions between gas particles, allowing the gas to adhere more closely to the principles of the kinetic molecular theory of gases. (A good way to remember this is to think of summertime, when the temperature is high and the pressure is low.)

17. **1** Heat energy always flows from an object with a higher temperature to an object with a lower temperature. The condition we know as "cold" is merely the absence of heat. Fast-moving particles in the hotter sample collide with the slow-moving particles in the colder sample. Energy is transferred as a result of these collisions.

18. **2** Equal volumes of gas at the same conditions of temperature and pressure contain the same number of particles. This relationship is known as Avogadro's Law. For example, the molar volume of a gas at STP = 22.4 L/mol no matter what type of gas you are working with.

19. **2** Water molecules are attracted to each other by **hydrogen bonding**. Hydrogen bonding is an unusually strong intermolecular force that occurs when hydrogen is bonded to a small, highly electronegative atom, such as F, O, or N. Molecules of H_2S do not contain F, O, or N atoms and will be attracted by weaker intermolecular forces. Water's strong hydrogen bonds explain why water is in the liquid state at room temperature, whereas H_2S is a gas at room temperature.

20. **3** Systems naturally tend to undergo changes that produce lower total energy and greater disorder or randomness (entropy). For example, during a landslide, the debris moves downslope toward a position of lower potential energy and eventually comes to rest in a random, chaotic mess.

21. **3** Amines, amides, and amino acids all contain the element carbon and are different categories of organic compounds. Reference Table R provides a list of organic functional groups. Amines and amides are on this table. Although they are not included on Reference Table R, amino acids are compounds that contain both the amine functional group and the organic acid functional group.

22. **2** Reference Table Q shows the general formulas for the three homologous series of hydrocarbons, known as the alkanes, alkenes, and alkynes. From this table, we see that only alkenes and alkynes contain multiple covalent bonds (double and triple bonds, respectively). C_2H_4 matches the general formula for alkenes, C_nH_{2n}.

WRONG CHOICES EXPLAINED:

(1), (3), and (4) These compounds are alkanes, since they match the general formula for alkanes, C_nH_{2n+2}.

23. **4** In saponification reactions, a fat reacts with a base to produce glycerol and soap. Review the following example:

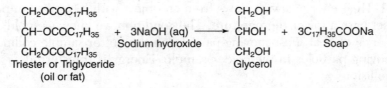

(NOTE: You need to be able to recognize seven major types of organic reactions for this exam. These include addition, substitution, esterification, polymerization, fermentation, saponification, and combustion. Memorize these reactions!)

WRONG CHOICES EXPLAINED:

(1) Polymerization reactions create long chains from many simple, identical molecules called monomers.

(2) Combustion reactions occur when fuel (usually a hydrocarbon) reacts with oxygen to produce carbon dioxide and water.

(3) Fermentation reactions involve the conversion of glucose into ethanol in the presence of an enzyme catalyst.

24. **3** Stresses can include changes in concentration, pressure, or temperature. Le Châtelier's principle states that chemical systems that are at equilibrium and experience a stress will respond to restore equilibrium.

25. **2** Cu atoms must lose two electrons to become Cu^{2+} ions. Oxidation numbers are assigned to elements to reflect the loss or gain of electrons.

WRONG CHOICES EXPLAINED:

(1), (3), and (4) The mass number is the sum of protons and neutrons in the nucleus of an atom. Protons and neutrons are not involved in the formation of ions.

26. **1** The mnemonic device "An Ox, Red Cat" is helpful for remembering that oxidation (loss of electrons) takes place at the anode, while reduction (gain of electrons) happens at the cathode.

27. **4** The spontaneous emission of beta particles is a form of radioactive decay. Beta decay occurs when the ratio of neutrons to protons in the nucleus of an atom is too large, causing the nucleus to become unstable.

28. **1** Notations for particles involved in nuclear chemistry are listed in Reference Table *O*. The only particle that is correctly paired with its notation is gamma radiation.

WRONG CHOICES EXPLAINED:
(2) $_2^4$He is an alpha particle, not a proton.
(3) $_{-1}^0\beta$ is a beta particle, not a neutron.
(4) $_1^1$H is a proton, not an alpha particle.

29. **2** Nuclear fusion produces energy by converting a small amount of mass into energy. The amount of energy produced can be calculated by Albert Einstein's famous equation $E = mc^2$, in which "m" is the mass lost in the fusion reaction and "c" is the speed of light.

30. **3** Radioactive dating makes use of the consistent half-life of radioisotopes to calculate the time elapsed during a certain event.

PART B–1

31. **1** Ions form when electrons enter or leave an atom, creating an imbalance between the number of protons and the number of electrons. Protons have a charge of 1+, and electrons have a charge of 1–. Since the ion described in this question has one more proton than the number of electrons, this ion will have a charge of 1+.

32. **3** The first electron shell (also known as a principal energy level) can only hold 2 electrons. The second electron shell can hold up to 8 electrons. Electrons will always seek the lowest energy positions available within the atom, which are positions near the nucleus. An electron configuration of 2-5-1 represents an electron in an excited state because it shows that an electron occupies the third electron shell before the second electron shell is filled.

33. **2** Reference Table *S* lists the melting points of the majority of elements on the Periodic Table. An element will be in the liquid phase if it is at a temperature above its melting point and below its boiling point. Aluminum would be a liquid at 1000. K since its melting point is 933 K.

34. **1** Ammonium nitrate is an ionic compound made from two oppositely charged polyatomic ions. The trick is to remember to consult Reference Table E to find the formulas and charges of these ions. The ammonium ion is NH_4^+, and the nitrate ion is NO_3^-. Ionic compounds are neutral in charge, so the formula for ammonium nitrate is NH_4NO_3 because the 1+ charge on the ammonium ion exactly cancels the 1− charge on the nitrate ion.

35. **1** Butene belongs to the homologous series of hydrocarbons known as the alkenes. Alkenes are one of the three homologous series of hydrocarbons listed on Reference Table Q. Alkenes have the general formula C_nH_{2n}. Reference Table P shows that the prefix "but-" means that a hydrocarbon has four carbon atoms, so butene has the molecular formula C_4H_8. However, this question asks for the *empirical formula* of butene, which is a simplified formula that is reduced to the lowest whole-number ratio. The empirical formula for butene, C_4H_8, is CH_2.

WRONG CHOICES EXPLAINED:
(2) C_2H_4 can be further simplified to CH_2.
(3) C_4H_6 is the formula for butyne, not butene.
(4) C_4H_8 is the molecular formula for butene.

36. **4** Conservation of charge means that the number of electrons that are lost through the process of oxidation equals the number of electrons that are gained through the process of reduction. To verify conservation of charge, take the sum of the charges on the reactant side and on the product side of the reaction. The sum of the charges should be the same on both sides. For the reaction displayed in choice (4):

$$3Fe^{3+} + Al \rightarrow 3Fe^{2+} + Al^{3+}$$
$$3(3+) + 0 = 3(2+) + (3+)$$
$$9+ = 9+$$

WRONG CHOICES EXPLAINED:
(1) The sum of the charges on the reactant side is 6+, while the sum of the charges on the product side is 7+.
(2) The sum of the charges on the reactant side is 6+, while the sum of the charges on the product side is 12+.
(3) The sum of the charges on the reactant side is 9+, while the sum of the charges on the product side is 10+.

37. **4** Synthesis reactions follow the general pattern A + B → AB. Two reactants combine to form one product.

WRONG CHOICES EXPLAINED:
(1) Decomposition reactions follow the general pattern AB → A + B. Decomposition reactions are the *opposite* of synthesis reactions.
(2) Double replacement reactions follow the general pattern AB + CD → AD + CB. Ions in two different ionic compounds switch places.
(3) Single replacement reactions follow the general pattern A + BC → B + AC. Uncombined elements appear on *both* sides of the equation.

38. **2** An Mg^{2+} ion must gain 2 electrons to become a neutral Mg atom. The Mg^{2+} ion has an electron configuration of 2-8. The Mg atom has an electron configuration of 2-8-2. Since the Mg atom has electrons in the third electron shell, it will have a larger radius than the Mg^{2+} ion, which does not have any electrons in the third electron shell.

39. **1** The least polar bond will have the lowest electronegativity difference between the atoms involved in the bond. Electronegativity values are found on Reference Table S. The electronegativity of H is 2.2, and the electronegativity of I is 2.7. The electronegativity difference of 0.5 (since 2.7 − 2.2 = 0.5) in the compound HI is the lowest of all the compounds offered as answer choices for this question.

40. **2** Look at Reference Table T to find the formula for parts per million:

$$\text{parts per million} = \frac{\text{mass of solute}}{\text{mass of solution}} \times 1,000,00$$

$$\text{parts per million} = \frac{0.125\,g}{250.\,g} \times 1,000,000 = 500.\ \text{ppm} = 5.00 \times 10^2\,\text{ppm}$$

41. **3** Use the heat equation for temperature changes from Reference Table T:

$$q = mC\Delta T$$

where q = heat, m = mass, C = specific heat capacity, and ΔT = change in temperature
   Plugging in the values given in the problem, and consulting Reference Table B for the specific heat capacity of water, results in:
   $q = 49.5$ g $(4.18\ \text{J/g} \cdot \text{°C})(66\text{°C} - 22\text{°C}) = 9104$ J $= 9.1 \times 10^3$ J (when expressed to two significant figures)
   (NOTE: The specific heat capacity of water can be written as either $4.18\ \text{J/g} \cdot \text{°C}$ or as $4.18\ \text{J/g} \cdot \text{K}$, since the interval of one Celsius degree equals the interval of one Kelvin degree.)

42. **2** Use the Combined Gas Law equation from Reference Table T to solve this problem:

$$\frac{P_1 V_1}{T_1} = \frac{P_2 V_2}{P_2}$$

Reference Table A provides the values for standard temperature and pressure.

Substituting the values from Reference Table A plus the information from the question yields:

$$\frac{(101.3\ \text{kPa})(11.2\,\text{L})}{273\,\text{K}} = \frac{(202.6\,\text{kPa})(V_2)}{300.\,\text{K}}$$

Rearranging this equation leads to:

$$V_2 = \frac{(101.3\,\text{kPa})(11.2\,\text{L})(300.\,\text{K})}{(273\,\text{K})(202.6\,\text{kPa})} = 6.15\,\text{L}$$

43. **2** For a chemical reaction to occur, molecules must collide with sufficient energy and proper orientation. Consider the following collision of reactant molecules:

This orientation immediately leads to the activated complex, which is a necessary step on the way to forming the products. None of the other orientations shown in the remaining choices immediately lead to the activated complex.

44. **1** The compound shown contains 4 carbon atoms and 10 hydrogen atoms, so its molecular formula is C_4H_{10}. This formula follows the general pattern for the alkane series of hydrocarbons, C_nH_{2n+2}, as shown on Reference Table Q. Reference Table P shows that the prefix for a hydrocarbon with 4 carbon atoms is *but-*. Putting together this prefix with the suffix for the alkane series, *-ane*, leads to the name "butane."

WRONG CHOICES EXPLAINED:
(2) Butene is an alkene with a formula of C_4H_8.
(3), (4) These hydrocarbons contain 10 carbon atoms.

45. **4** The actual value for the density of vanadium, 6.0 g/cm^3, is listed on Reference Table S. Use the percent error formula on Reference Table T to set up the calculation for percent error:

$$\% \text{ error} = \frac{\text{measured value} - \text{accepted value}}{\text{accepted value}} \times 100$$

Plugging in the values from the question:

$$\% \text{ error} = \frac{6.9 \text{ g/cm}^3 - 6.0 \text{ g/cm}^3}{6.0 \text{ g/cm}^3} \times 100 = 15\%$$

46. **3** Oxidation is the loss of electrons. As a reactant, cadmium is a pure element and has an oxidation number of 0. In the compound Cd(OH)$_2$, on the product side of the equation, each Cd atom has an oxidation number of 2+. Cadmium atoms must lose two electrons to go from an oxidation number of 0 to an oxidation number of 2+.

WRONG CHOICES EXPLAINED:

(1) This half-reaction is an example of *reduction*, which is the gain of electrons.

(2) This oxidation half-reaction is not balanced for charge [4+ ≠ (2+) + (2−)].

(4) This reduction half-reaction is not balanced for charge [0 + (2−) ≠ (2+)].

47. **4** Reference Table J lists the most active metals at the top of the table down to the least active metals at the bottom of the table. Pure metal atoms never gain electrons, but metal ions can gain back the electrons they previously lost in the presence of a more active metal. Ni^{2+} ions will gain electrons back in the presence of zinc (Zn), which is higher on Reference Table J and is therefore a more active metal.

WRONG CHOICES EXPLAINED:

(1), (2), and (3) These metals are found below nickel on Reference Table J, signifying that they have less ability to oxidize than nickel does and consequently they will not be able to push electrons back onto the Ni^{2+} ion.

48. **3** Electrolytes are substances that dissolve in water to form aqueous solutions that are capable of conducting electricity. The more ions there are in a solution, the more electrically conductive the solution becomes.

49. **2** The pH is the negative of the exponent of the base 10 concentration of the hydrogen (or hydronium) ion. Using this definition, a solution with a pH of 7.0 has a hydronium ion concentration of 10^{-7} M, while a solution with a pH of 5.0 has a hydronium ion concentration of 10^{-5} M. The solution with a pH of 5.0 has 100 times more hydronium ions than the solution with a pH of 7.0.

50. **4** The term "transmutation" refers to any change in the nucleus of an atom that converts it from one element to another. The nuclear reaction in this question shows a collision, between two elements with smaller nuclei, that produces an element with a larger nucleus.

WRONG CHOICES EXPLAINED:

(1) Neutralization reactions occur when an acid reacts with a base to form water and a salt.

(2) Polymerization reactions are organic reactions in which small molecules (monomers) combine to make very long chains (polymers).

(3) Substitution reactions are organic reactions in which an atom (usually a halogen) substitutes for a hydrogen in an organic compound.

PART B–2

[All questions in Part B–2 are worth 1 point.]

51. To find the percent composition by mass of oxygen in Al_2O_3, use the percent composition formula found on Reference Table *T*:

$$\%\text{composition by mass} = \frac{\text{mass of part}}{\text{mass of whole}} \times 100$$

where the "part" is the mass of the oxygen in the compound, and the "whole" is the gram-formula mass of the entire compound

$$\%\text{ composition by mass of oxygen in } Al_2O_3 = \frac{3(16.0\,\text{g/mol})}{102.0\,\text{g/mol}} \times 100$$

(NOTE: It is not necessary to solve for the percent composition by mass of oxygen in this question since you are only asked for the *numerical setup*.)

52. **Distillation** is the laboratory process that is used to separate liquid mixtures based on boiling points. As the liquid mixture begins to boil, the component of the mixture with the lower boiling point will boil away first. By passing the vapors through a condenser, the components of the mixture can be collected separately.

53. All three isotopes have the **same number of protons**. The number of protons is equal to the atomic number of the element. All three isotopes have 12 protons and are different isotopes of the element magnesium.

54. Valence electrons are located in the outermost energy level of an atom. Valence electrons are important because they are the electrons that are involved in chemical bonding. Isotope *D* contains 12 protons, so it represents the element magnesium. Look at the Periodic Table in your Reference Tables to find the electron configuration for magnesium, which is 2-8-**2**. The outermost energy level contains **2** electrons, which are the valence electrons for this element.

55. Electron shells that are closer to the nucleus have lower energy than electron shells that are farther from the nucleus. This is because negatively charged electrons are attracted to the positively charged nucleus. **An electron in the first electron shell of any atom will have less energy than an electron in the second electron shell.**

56. **Electronegativity decreases** as the atomic number increases in all groups on the Periodic Table, including the metals in Group 2. As the atomic number increases going down a group of elements on the Periodic Table, the valence (outermost) electrons are located in electron shells that are farther from the nucleus. The nucleus of an atom has far less force of attraction for new electrons when the new electrons will be added to a more distant electron shell.

57. **The elements in Group 2 on the Periodic Table each have 2 valence electrons.** Elements with the same number of valence electrons will have similar chemical properties. Valence electrons are the electrons involved in bonding, and they have a large influence on the stability of an element.

58. The first ionization energy is the energy that is needed to remove the outermost, or highest energy, electron from an atom. **Barium's valence electrons are in the 6th electron shell, whereas magnesium's valence electrons are in the 3rd electron shell. Less energy is required to remove electrons that are farther from the nucleus.**

59. Refer to Reference Table *G* to answer this question. The table shows that approximately **16 g** of SO_2 would be dissolved in the solution described in this question.
   (NOTE: All values from **15 g to 18 g** were accepted as correct answers to this question.)

60. From Reference Table *G* it is clear that, **as the temperature increases, the solubility of SO₂ decreases**. SO_2 is a gas, and gases at higher temperatures have sufficient kinetic energy to overcome the intermolecular forces with solvent molecules.

61. **The solubility of SO₂ gas increases at higher pressures** because the increased pressure forces more SO_2 molecules into the solution, as shown in the following diagram:

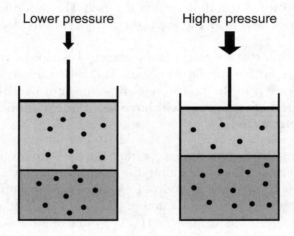

62. The information supplied for this question states that a solid sample is heated continuously until it becomes a gas. The interval *AB* represents the solid sample before it melts. Particle diagrams of solids need to show high density, fixed size and shape, and regular arrangement of particles, as shown in the following diagram. Make sure to draw *at least six* particles in your diagram, as directed by the question.

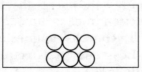

63. Temperature is directly related to the average kinetic energy of particles in a sample of matter. **Molecules in the sample have greater average kinetic energy during interval *DE*** because the sample is at a higher temperature during this period of time.

64. Phase changes occur during the "plateaus," or flat intervals, on a heating curve. The temperature remains constant during phase changes. Boiling happens when particles are transitioning from the liquid phase to the gas phase. This transition occurs during interval *DE* on the heating curve. The directions state that you must mark an X on the axis labeled "Temperature (°C)," so your "X" needs to be drawn directly in line with the boiling plateau, *DE*, as shown in the following diagram:

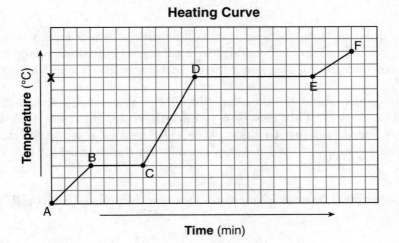

Heating Curve

65. As heating proceeds along this heating curve, the particles in the sample move faster and spread farther apart, but the molecules in the sample are the very same substance throughout the entire heating curve. This represents a physical change of matter, since **no new substance is formed**. On the other hand, chemical changes involve the formation of new substances with different properties.

PART C

[All questions in Part C are worth 1 point.]

66. The condition known as dynamic equilibrium occurs when **the rate of the forward reaction equals the rate of the reverse reaction**. One real life example of dynamic equilibrium is traffic on a street moving in opposite directions at an equal rate.

67. Lewis electron-dot diagrams are drawn to show the distribution of valence electrons within a molecule. These diagrams help to illustrate the number and types of bonds in a compound. The following illustrations are examples of a Lewis electron-dot diagram for water. Note that, in the finished diagram, the oxygen atom achieves a stable octet of valence electrons by sharing each hydrogen's single valence electron.

$$H—\overset{..}{\underset{..}{O}}—H \quad OR \quad H\overset{..}{\underset{..}{O}}H$$

68. In order to react, particles must collide with sufficient energy (the activation energy) and proper orientation. **More effective collisions occur** between water molecules and solid carbon if the carbon is spread out over a larger area.

69. Reference Table I shows that the heat of reaction (ΔH) for dissolving KNO_3 is +34.89 kJ. **This reaction is endothermic, meaning that heat is required for dissolving to take place.** The heat of reaction for dissolving KNO_3 will be supplied by the water, causing the temperature of the water to decrease.

70. The percent by mass (percent composition) formula is given on Reference Table T:

$$\% \text{ composition by mass} = \frac{\text{mass of part}}{\text{mass of whole}} \times 100$$

where the "part" is the mass of the KNO_3 in the solution and the "whole" is the mass of the entire solution

$$\% \text{ composition by mass of } KNO_3 = \frac{5.0\,\text{grams}}{15.0\,\text{grams}} \times 100 = \mathbf{33.3\%}$$

(NOTE: The mass of the solution (15.0 grams) equals the mass of KNO_3 in mixture 2 (5.0 grams) added to the mass of the original water (10.0 grams). Also, note that all values from **33%** to **33.3%** were accepted as correct answers to this question.)

71. Dissolved ions lower the freezing point of a solution. This is the colligative property known as freezing point depression. The more ions that are dissolved, the lower the freezing point. **The freezing point of mixture 4 will be lower than the freezing point of water** due to the dissolved KNO_3.

72. The heat of reaction is defined as the difference between the potential energy of the products (present after the reaction) and the potential energy of the reactants (present before the reaction). The heat of reaction is indicated on a potential energy diagram as follows:

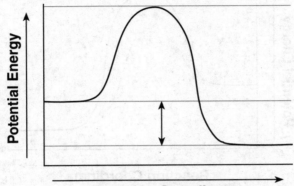

(NOTE: As per the question, the arrow **must** be a double-headed arrow.)

73. All coefficients in chemical reactions represent particles or moles. Mole ratios can be used to find the number of moles of any substance in a cahemical equation if the moles of another substance are known. Two methods of setting up this problem are:

$$\frac{8.0 \text{ mol CO}}{1} \times \frac{1 \text{ mol O}_2}{2 \text{ mol CO}} = \textbf{4.0 mol O}_2$$

You could also set up mole ratios:

$$\frac{8.0 \text{ mol CO}}{2} = \frac{x \text{ mol O}_2}{1}$$

$$x = \textbf{4.0 mol O}_2$$

74. Catalysts speed up chemical reactions by *lowering* the activation energy (E_a) for the reaction. With a lower E_a, less energy is needed for effective collisions. You can show the lowering of the E_a by drawing the dashed line as follows:

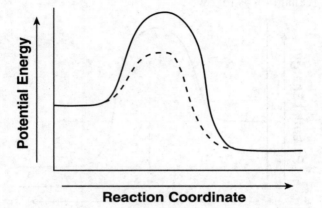

75. Compound *A* contains the –OH functional group. According to Reference Table *R*, compounds with an –OH functional group are classified as **alcohols**.

76. Organic chemistry is the study of compounds that contain the element **carbon**.

77. Hydrocarbons are organic compounds that contain only two elements: carbon and hydrogen. **Since compound *B* also contains oxygen**, it cannot be classified as a hydrocarbon.

78. According to the Arrhenius theory of acids and bases, **H$^+$** (the hydrogen ion) is the only positive ion present in acidic solutions.
   (NOTE: Other acceptable answers include **H$_3$O$^+$**, or **the hydronium ion**.)

79. The completed equation should read as follows:

$$2NaOH(aq) + H_2SO_4(aq) \rightarrow 2H_2O(\ell) + \mathbf{Na_2SO_4}(aq)$$

The general equation for a neutralization equation is:

$$\text{Acid} + \text{Base} \rightarrow \text{Water} + \text{Salt}$$

Note that the salt will be an ionic compound. The Law of Conservation of Mass states that all atoms that enter a chemical reaction must leave the reaction in equal numbers. A neutralization equation is a specific type of double replacement reaction in which the ions switch places and leave with the oppositely charged ion on the other compound. The sodium ion (Na^+) switches places with the hydrogen ion (H^+) to form the ionic compound sodium sulfate (Na_2SO_4).

80. On the pH scale, which ranges from 0 to 14, a pH of 7 is exactly neutral. This is the pH of pure water. The graph shows that a pH of 7 is reached after the addition of **25 mL of NaOH**.

(NOTE: All values from **24 mL to 26 mL** were accepted as correct answers to this question.)

81. The graph shows that after 50. mL of NaOH(aq) were added, the solution reached a pH of approximately 12.5. Reference Table *M* lists the colors of common acid-base indicators at different pH ranges. Phenolphthalein will be **pink** when the pH is greater than 9.

82. Fission is the splitting of a large atomic nucleus (usually U-235) into two smaller nuclei. The fission reaction is initiated when the target nucleus is struck by a neutron, and the resulting emission of neutrons as products makes a chain reaction possible. The nuclear equation in this question represents a fission reaction because **the $^{235}_{92}U$ nuclide is bombarded by a neutron and splits into two smaller nuclides**.

83. The isotopic notation for neodymium-150, $^{150}_{60}Nd$, shows that Nd-150 has an atomic number of 60 and, therefore, this element has **60 protons**. The neutrons in this isotope of neodymium can be found by subtracting the atomic number (the number of protons) from the mass number of 150, which is the sum of the protons and the neutrons. Thus, this element has **90 neutrons**, since $150 - 60 = 90$ neutrons.

84. The completed equation should read as follows:

$$^{81}_{32}Ge \rightarrow \, ^{0}_{-1}e + \, ^{81}_{33}As$$

When balancing a nuclear equation, the sum of the atomic numbers (the bottom numbers) and the mass numbers (the top numbers) must be equal on both sides of the arrow.

85. One half-life is the time required for a radioisotope to decay to half of its original mass. To find the total number of half-lives required for Ge-81 to decay from a mass of 16.00 grams to a mass of 1.00 gram, divide the mass in half over a series of steps.

In the first half-life, Ge-81 decays from 16.00 grams to 8.00 grams. (The other 8.00 grams undergo radioactive decay). In the second half-life, the 8.00 grams of Ge-81 remaining decay to 4.00 grams. In the third half-life, the 4.00 grams of Ge-81 remaining decay to 2.00 grams. In the fourth half-life, the 2.00 grams of Ge-81 remaining decay to 1.00 gram. Thus, four half-lives are needed for the radioactive decay of Ge-81 from 16.00 grams to 1.00 gram. The information given for questions 82–85 states that Ge-81 has a half-life of 7.6 seconds. Therefore, the time required for four half-lives is 4×7.6 seconds = **30.4 seconds**.

Mark (✓) the questions you answered correctly. Count the number of checks and follow the formulas given to determine your score on each topic.

Core Area   ☐ Questions Answered Correctly

45: (1)

Section M—Math Skills
☐ Number of checks ÷ 1 × 100 = ___ %

33: (1)

Section R—Reading Skills
☐ Number of checks ÷ 1 × 100 = ___ %

1–4, 31, 32, 53–55, 83: (10)

Section I—Atomic Concepts
☐ Number of checks ÷ 10 × 100 = ___ %

5–7, 56–58: (6)

Section II—Periodic Table
☐ Number of checks ÷ 6 × 100 = ___ %

8, 34–37, 51, 73: (7)

Section III—Moles/Stoichiometry
☐ Number of checks ÷ 7 × 100 = ___ %

9–13, 15, 38, 39, 67: (9)

Section IV—Chemical Bonding
☐ Number of checks ÷ 9 × 100 = ___ %

14, 16–19, 40–42, 52, 59–65, 70, 71: (18)

Section V—Physical Behavior of Matter
☐ Number of checks ÷ 18 × 100 = ___ %

20, 24, 43, 66, 68, 69, 72, 74: (8)

Section VI—Kinetics and Equilibrium
☐ Number of checks ÷ 8 × 100 = ___ %

Core Area	☐ Questions Answered Correctly

21–23, 44, 75–77: (7)

Section VII—Organic Chemistry
☐ Number of checks ÷ 7 × 100 = ___ %

25, 26, 46, 47: (4)

Section VIII—Oxidation–Reduction
☐ Number of checks ÷ 4 × 100 = ___ %

48, 49, 78–81: (6)

Section IX—Acids, Bases, and Salts
☐ Number of checks ÷ 6 × 100 = ___ %

27–30, 50, 82, 84, 85: (8)

Section X—Nuclear Chemistry
☐ Number of checks ÷ 8 × 100 = ___ %

Examination June 2019

Chemistry—Physical Setting

PART A

Answer all questions in this part.

Directions (1–30): For *each* statement or question, write in the answer space the *number* of the word or expression that, of those given, best completes the statement or answers the question. Some questions may require the use of the *2011 Edition Reference Tables for Physical Setting/Chemistry*.

1 Which particles are found in the nucleus of an argon atom?

  (1) protons and electrons
  (2) positrons and neutrons
  (3) protons and neutrons
  (4) positrons and electrons   1 _____

2 The diagram below represents a particle traveling through an electric field.

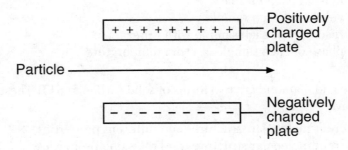

An electric field exists between the two plates.

Which particle remains undeflected when passing through this electric field?

(1) proton   (3) neutron

(2) electron   (4) positron   2 _____

3 The mass of an electron is

(1) equal to the mass of a proton
(2) equal to the mass of a neutron
(3) greater than the mass of a proton
(4) less than the mass of a neutron   3 _____

4 Compared to the energy of an electron in the second shell of an atom of sulfur, the energy of an electron in the

(1) first shell is lower
(2) first shell is the same
(3) third shell is lower
(4) third shell is the same   4 _____

5 In the ground state, an atom of which element has seven valence electrons?

(1) sodium   (3) nitrogen

(2) phosphorus   (4) fluorine   5 _____

6 Which information is sufficient to differentiate a sample of sodium from a sample of silver?

(1) the mass of each sample
(2) the volume of each sample
(3) the reactivity of each sample with water
(4) the phase of each sample at room temperature   6 _____

7 Graphite and diamond are two forms of solid carbon at STP. These forms have

(1) different molecular structures and different properties
(2) different molecular structures and the same properties
(3) the same molecular structures and different properties
(4) the same molecular structures and the same properties   7 _____

8 As the first five elements in Group 14 are considered in order from top to bottom, there are changes in both the

(1) number of valence shell electrons and number of first shell electrons

(2) electronegativity values and number of first shell electrons

(3) number of valence shell electrons and atomic radii

(4) electronegativity values and atomic radii   8 _____

9 Which statement explains why NaBr is classified as a compound?

(1) Na and Br are chemically combined in a fixed proportion.

(2) Na and Br are both nonmetals.

(3) NaBr is a solid at 298 K and standard pressure.

(4) NaBr dissolves in H_2O at 298 K.   9 _____

10 Which two terms represent types of chemical formulas?

(1) fission and fusion

(2) oxidation and reduction

(3) empirical and structural

(4) endothermic and exothermic   10 _____

11 During all chemical reactions, charge, mass, and energy are

(1) condensed   (3) decayed

(2) conserved   (4) decomposed   11 _____

12 The degree of polarity of a covalent bond between two atoms is determined by calculating the difference in their

(1) atomic radii   (3) electronegativities

(2) melting points   (4) ionization energies   12 _____

13 Which substance can *not* be broken down by a chemical change?

(1) ammonia   (3) methane

(2) magnesium   (4) water   13 _____

14 Which statement describes the components of a mixture?

   (1) Each component gains new properties.
   (2) Each component loses its original properties.
   (3) The proportions of components can vary.
   (4) The proportions of components cannot vary.   14 _____

15 Table sugar can be separated from a mixture of table sugar and sand at STP by adding

   (1) sand, stirring, and distilling at 100.°C
   (2) sand, stirring, and filtering
   (3) water, stirring, and distilling at 100.°C
   (4) water, stirring, and filtering   15 _____

16 Which statement describes the particles of an ideal gas, based on the kinetic molecular theory?

   (1) The volume of the particles is considered negligible.
   (2) The force of attraction between the particles is strong.
   (3) The particles are closely packed in a regular, repeating pattern.
   (4) The particles are separated by small distances, relative to their size.   16 _____

17 During which two processes does a substance release energy?

   (1) freezing and condensation
   (2) freezing and melting
   (3) evaporation and condensation
   (4) evaporation and melting   17 _____

18 Based on Table I, which compound dissolves in water by an exothermic process?

   (1) NaCl   (3) NH_4Cl
   (2) NaOH   (4) NH_4NO_3 18 _____

19 At STP, which property of a molecular substance is determined by the arrangement of its molecules?

  (1) half-life
  (2) molar mass
  (3) physical state
  (4) percent composition   19 _____

20 Equilibrium can be reached by

  (1) physical changes, only
  (2) nuclear changes, only
  (3) both physical changes and chemical changes
  (4) both nuclear changes and chemical changes   20 _____

21 Which value is defined as the difference between the potential energy of the products and the potential energy of the reactants during a chemical change?

  (1) heat of fusion
  (2) heat of reaction
  (3) heat of deposition
  (4) heat of vaporization   21 _____

22 The effect of a catalyst on a chemical reaction is to provide a new reaction pathway that results in a different

  (1) potential energy of the products
  (2) heat of reaction
  (3) potential energy of the reactants
  (4) activation energy   22 _____

23 Chemical systems in nature tend to undergo changes toward

  (1) lower energy and lower entropy
  (2) lower energy and higher entropy
  (3) higher energy and lower entropy
  (4) higher energy and higher entropy   23 _____

24 The atoms of which element bond to one another in chains, rings, and networks?

  (1) barium   (3) iodine
  (2) carbon   (4) mercury   24 _____

25 What is the general formula for the homologous series that includes ethene?

(1) C_nH_{2n}

(2) C_nH_{2n-6}

(3) C_nH_{2n-2}

(4) C_nH_{2n+2} 25 _____

26 When an F atom becomes an F^- ion, the F atom

(1) gains a proton

(2) loses a proton

(3) gains an electron

(4) loses an electron   26 _____

27 Which substance is an Arrhenius base?

(1) HNO_3

(2) H_2SO_3

(3) $Ca(OH)_2$

(4) CH_3COOH 27 _____

28 In which type of nuclear reaction do two light nuclei combine to produce a heavier nucleus?

(1) positron emission

(2) gamma emission

(3) fission

(4) fusion   28 _____

29 Using equal masses of reactants, which statement describes the relative amounts of energy released during a chemical reaction and a nuclear reaction?

(1) The chemical and nuclear reactions release equal amounts of energy.

(2) The nuclear reaction releases half the amount of energy of the chemical reaction.

(3) The chemical reaction releases more energy than the nuclear reaction.

(4) The nuclear reaction releases more energy than the chemical reaction.   29 _____

30 The ratio of the mass of U-238 to the mass of Pb-206 can be used to

(1) diagnose thyroid disorders

(2) diagnose kidney function

(3) date geological formations

(4) date once-living things   30 _____

PART B–1

Answer all questions in this part.

Directions (31–50): For *each* statement or question, write in the answer space the *number* of the word or expression that, of those given, best completes the statement or answers the question. Some questions may require the use of the *2011 Edition Reference Tables for Physical Setting/Chemistry.*

31 The bright-line spectra of four elements, G, J, L, and M, and a mixture of *at least two* of these elements is given below.

Bright-Line Spectra

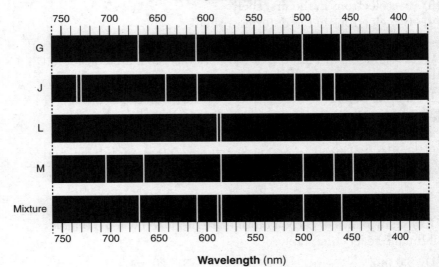

Wavelength (nm)

Which elements are present in this mixture?

(1) G and J

(2) G and L

(3) M, J, and G

(4) M, J, and L   31 _____

32 Which electron configuration represents an atom of chlorine in an excited state?

(1) 2-8-7-2

(2) 2-8-7

(3) 2-8-8

(4) 2-7-8   32 _____

33 A student measures the mass and volume of a sample of aluminum at room temperature, and calculates the density of Al to be 2.85 grams per cubic centimeter. Based on Table S, what is the percent error for the student's calculated density of Al?

(1) 2.7%   (3) 5.6%

(2) 5.3%   (4) 95%   33 _____

34 Magnesium and calcium have similar chemical properties because their atoms in the ground state have

(1) equal numbers of protons and electrons

(2) equal numbers of protons and neutrons

(3) two electrons in the first shell

(4) two electrons in the outermost shell   34 _____

35 As the elements in Period 2 of the Periodic Table are considered in order from left to right, which property generally *decreases*?

(1) atomic radius   (3) ionization energy

(2) electronegativity   (4) nuclear charge   35 _____

36 Given the balanced equation for the reaction of butane and oxygen:

$$2C_4H_{10} + 13O_2 \rightarrow 8CO_2 + 10H_2O + energy$$

How many moles of carbon dioxide are produced when 5.0 moles of butane react completely?

(1) 5.0 mol   (3) 20. mol

(2) 10. mol   (4) 40. mol   36 _____

37 What is the percent composition by mass of nitrogen in the compound N_2H_4 (gram-formula mass = 32 g/mol)?

(1) 13%   (3) 88%

(2) 44%   (4) 93%   37 _____

38 Which ion in the ground state has the same electron configuration as an atom of neon in the ground state?

(1) Ca^{2+} (3) Li^+

(2) Cl^- (4) O^{2-} 38 _____

39 The molar masses and boiling points at standard pressure for four
   compounds are given in the table below.

Compound	Molar Mass (g/mol)	Boiling Point (K)
HF	20.01	293
HCl	36.46	188
HBr	80.91	207
HI	127.91	237

   Which compound has the strongest intermolecular forces?

   (1) HF   (3) HBr
   (2) HCl   (4) HI   39 _____

40 Which particle model diagram represents xenon at STP?

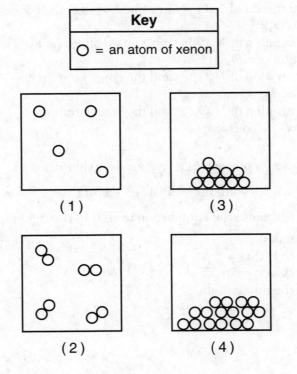

41 What is the amount of heat absorbed when the temperature of 75 grams of water increases from 20.°C to 35°C?

(1) 1100 J

(3) 6300 J

(2) 4700 J

(4) 11 000 J

41 _____

42 Which sample of HCl(aq) reacts at the fastest rate with a 1.0-gram sample of iron filings?

(1) 10. mL of 1 M HCl(aq) at 10.°C

(2) 10. mL of 1 M HCl(aq) at 25°C

(3) 10. mL of 3 M HCl(aq) at 10.°C

(4) 10. mL of 3 M HCl(aq) at 25°C

42 _____

43 Given the equation representing a system at equilibrium:

$$N_2O_4(g) \rightleftharpoons 2NO_2(g)$$

Which statement describes the concentration of the two gases in this system?

(1) The concentration of $N_2O_4(g)$ must be less than the concentration of $NO_2(g)$.

(2) The concentration of $N_2O_4(g)$ must be greater than the concentration of $NO_2(g)$.

(3) The concentration of $N_2O_4(g)$ and the concentration of $NO_2(g)$ must be equal.

(4) The concentration of $N_2O_4(g)$ and the concentration of $NO_2(g)$ must be constant.

43 _____

44 Given the equation representing a system at equilibrium:

$$PCl_5(g) + energy \rightleftharpoons PCl_3(g) + Cl_2(g)$$

Which change will cause the equilibrium to shift to the right?

(1) adding a catalyst

(2) adding more $PCl_3(g)$

(3) increasing the pressure

(4) increasing the temperature

44 _____

45 Given the formula representing a molecule:

$$H-\underset{\underset{\displaystyle H}{|}}{\overset{\overset{\displaystyle H}{|}}{C}}-\underset{\underset{\displaystyle H}{|}}{\overset{\overset{\displaystyle H}{|}}{C}}-\underset{\underset{\displaystyle H}{|}}{\overset{\overset{\displaystyle H}{|}}{C}}-\underset{\underset{\displaystyle H}{|}}{\overset{\overset{\displaystyle H}{|}}{C}}-\underset{\underset{\displaystyle H}{|}}{\overset{\overset{\displaystyle H}{|}}{C}}-N\overset{\displaystyle H}{\underset{\displaystyle H}{}}$$

A chemical name for this compound is

(1) pentanone
(2) 1-pentanol

(3) 1-pentanamine
(4) pentanamide

45 _____

46 Given the formula of a compound:

$$H-C\equiv C-\underset{\underset{\displaystyle H}{|}}{\overset{\overset{\displaystyle H}{|}}{C}}-\underset{\underset{\displaystyle H}{|}}{\overset{\overset{\displaystyle H}{|}}{C}}-H$$

This compound is classified as an

(1) aldehyde
(2) alkene

(3) alkyne
(4) alcohol

46 _____

47 Which equation represents fermentation?

(1) $C_2H_4 + H_2O \rightarrow CH_3CH_2OH$
(2) $C_2H_4 + HCl \rightarrow CH_3CH_2Cl$
(3) $C_6H_{12}O_6 \rightarrow 2CH_3CH_2OH + 2CO_2$
(4) $2CH_3CHO \rightarrow C_3H_5CHO + H_2O$

47 _____

48 Given the equation representing a reaction:

$$3CuCl_2(aq) + 2Al(s) \rightarrow 3Cu(s) + 2AlCl_3(aq)$$

The oxidation number of copper changes from

(1) +1 to 0
(2) +2 to 0

(3) +2 to +1
(4) +6 to +3

48 _____

49 Given the equation representing a reversible reaction:

$$CH_3COOH(aq) + H_2O(\ell) \rightleftharpoons$$
$$CH_3COO^-(aq) + H_3O^+(aq)$$

According to one acid-base theory, the two H^+ donors in the equation are

(1) CH_3COOH and H_2O
(2) CH_3COOH and H_3O^+
(3) CH_3COO^- and H_2O
(4) CH_3COO^- and H_3O^+ 49 _____

50 Which nuclear equation represents a spontaneous decay?

(1) $^{222}_{86}Rn \rightarrow \, ^{218}_{84}Po + \, ^4_2He$

(2) $^{27}_{13}Al + \, ^4_2He \rightarrow \, ^{30}_{15}P + \, ^1_0n$

(3) $^{235}_{92}U + \, ^1_0n \rightarrow \, ^{139}_{56}Ba + \, ^{96}_{36}Kr + 3^1_0n$

(4) $^7_3Li + \, ^1_1H \rightarrow \, ^4_2He + \, ^4_2He$ 50 _____

PART B–2

Answer all questions in this part.

Directions (51–65): Record your answers on the answer sheet provided. Some questions may require the use of the *2011 Edition Reference Tables for Physical Setting/Chemistry*.

51 Draw a structural formula for methanal. [1]

Base your answers to questions 52 through 54 on the information below and on your knowledge of chemistry.

The atomic mass and natural abundance of the naturally occurring isotopes of hydrogen are shown in the table below.

Naturally Occurring Isotopes of Hydrogen

Isotope	Common Name of Isotope	Atomic Mass (u)	Natural Abundance (%)
H-1	protium	1.0078	99.9885
H-2	deuterium	2.0141	0.0115
H-3	tritium	3.0160	negligible

The isotope H-2, also called deuterium, is usually represented by the symbol "D." Heavy water forms when deuterium reacts with oxygen, producing molecules of D_2O.

52 Explain, in terms of subatomic particles, why atoms of H-1, H-2, and H-3 are each electrically neutral. [1]

53 Determine the formula mass of heavy water, D_2O. [1]

54 Based on Table *N*, identify the decay mode of tritium. [1]

Base your answers to questions 55 through 57 on the information below and on your knowledge of chemistry.

At 23°C, 85.0 grams of $NaNO_3(s)$ are dissolved in 100. grams of $H_2O(\ell)$.

55 Convert the temperature of the $NaNO_3(s)$ to kelvins.   [1]

56 Based on Table *G*, determine the additional mass of $NaNO_3(s)$ that must be dissolved to saturate the solution at 23°C.   [1]

57 State what happens to the boiling point and freezing point of the solution when the solution is diluted with an additional 100. grams of $H_2O(\ell)$.   [1]

Base your answers to questions 58 through 61 on the information below and on your knowledge of chemistry.

A 200.-milliliter sample of $CO_2(g)$ is placed in a sealed, rigid cylinder with a movable piston at 296 K and 101.3 kPa.

58 State a change in temperature and a change in pressure of the $CO_2(g)$ that would cause it to behave more like an ideal gas.   [1]

59 Determine the volume of the sample of $CO_2(g)$ if the temperature and pressure are changed to 336 K and 152.0 kPa.   [1]

60 State, in terms of *both* the frequency and force of collisions, what would result from decreasing the temperature of the original sample of $CO_2(g)$, at constant volume.   [1]

61 Compare the mass of the original 200.-milliliter sample of $CO_2(g)$ to the mass of the $CO_2(g)$ sample when the cylinder is adjusted to a volume of 100. milliliters.   [1]

Base your answers to questions 62 through 65 on the information below and on your knowledge of chemistry.

Cobalt-60 is an artificial isotope of Co-59. The incomplete equation for the decay of cobalt-60, including beta and gamma emissions, is shown below.

$$_{27}^{60}\text{Co} \rightarrow X + _{-1}^{0}\text{e} + _{0}^{0}\gamma$$

62 Explain, in terms of *both* protons and neutrons, why Co-59 and Co-60 are isotopes of cobalt. [1]

63 Compare the penetrating power of the beta and gamma emissions. [1]

64 Complete the nuclear equation, on your answer sheet, for the decay of cobalt-60 by writing a notation for the missing product. [1]

65 Based on Table N, determine the total time required for an 80.00-gram sample of cobalt-60 to decay until only 10.00 grams of the sample remain unchanged. [1]

PART C

Answer all questions in this part.

Directions (66–85): Record your answers on the answer sheet provided. Some questions may require the use of the *2011 Edition Reference Tables for Physical Setting/Chemistry*.

Base your answers to questions 66 through 69 on the information below and on your knowledge of chemistry.

During a laboratory activity, appropriate safety equipment was used and safety procedures were followed. A laboratory technician heated a sample of solid $KClO_3$ in a crucible to determine the percent composition by mass of oxygen in the compound. The unbalanced equation and the data for the decomposition of solid $KClO_3$ are shown below.

$$KClO_3(s) \rightarrow KCl(s) + O_2(g)$$

Lab Data and Calculated Results

Object or Material	Mass (g)
empty crucible and cover	22.14
empty crucible, cover, and $KClO_3$	24.21
$KClO_3$	2.07
crucible, cover, and KCl after heating	23.41
KCl	?
O_2	0.80

66 Write a chemical name for the compound that decomposed. [1]

67 Based on the lab data, show a numerical setup to determine the number of moles of O_2 produced. Use 32 g/mol as the gram-formula mass of O_2. [1]

68 Based on the lab data, determine the mass of KCl produced in the reaction. [1]

69 Balance the equation on your answer sheet for the decomposition of $KClO_3$, using the smallest whole-number coefficients. [1]

Base your answers to questions 70 through 73 on the information below and on your knowledge of chemistry.

A bottled water label lists the ions dissolved in the water. The table below lists the mass of some ions dissolved in a 500.-gram sample of the bottled water.

Ions in 500. g of Bottled Water

Ion Formula	Mass (g)
Ca^{2+}	0.040
Mg^{2+}	0.013
Na^+	0.0033
SO_4^{2-}	0.0063
HCO_3^-	0.180

70 State the number of significant figures used to express the mass of hydrogen carbonate ions in the table above. [1]

71 Based on Table F, write the formula of the ion in the bottled water table that would form the *least* soluble compound when combined with the sulfate ion. [1]

72 Show a numerical setup for calculating the parts per million of the Na^+ ions in the 500.-gram sample of the bottled water. [1]

73 Compare the radius of a Mg^{2+} ion to the radius of a Mg atom. [1]

Base your answers to questions 74 through 77 on the information below and on your knowledge of chemistry.

Ethyl ethanoate is used as a solvent for varnishes and in the manufacture of artificial leather. The formula below represents a molecule of ethyl ethanoate.

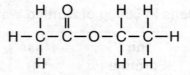

74 Identify the element in ethyl ethanoate that makes it an organic compound.   [1]

75 Write the empirical formula for this compound.   [1]

76 Write the name of the class of organic compounds to which this compound belongs.   [1]

77 Determine the number of electrons shared in the bond between a hydrogen atom and a carbon atom in the molecule.   [1]

Base your answers to questions 78 through 80 on the information below and on your knowledge of chemistry.

An operating voltaic cell has magnesium and silver electrodes. The cell and the ionic equation representing the reaction that occurs in the cell are shown below.

Voltaic Cell

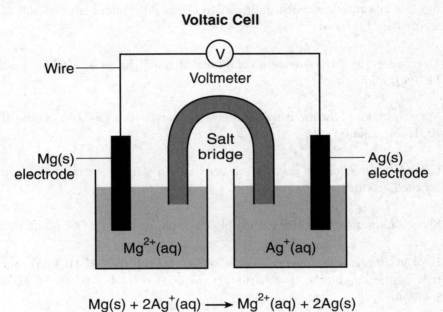

$$Mg(s) + 2Ag^+(aq) \longrightarrow Mg^{2+}(aq) + 2Ag(s)$$

78 State the purpose of the salt bridge in this cell.   [1]

79 Write a balanced equation for the half-reaction that occurs at the magnesium electrode in this cell.   [1]

80 Explain, in terms of electrical energy, how electrolysis reactions differ from voltaic cell reactions.   [1]

Base your answers to questions 81 through 85 on the information below and on your knowledge of chemistry.

In a laboratory investigation, an HCl(aq) solution with a pH value of 2 is used to determine the molarity of a KOH(aq) solution. A 7.5-milliliter sample of the KOH(aq) is exactly neutralized by 15.0 milliliters of the 0.010 M HCl(aq). During this laboratory activity, appropriate safety equipment is used and safety procedures are followed.

81 Determine the pH value of a solution that is ten times less acidic than the HCl(aq) solution.   [1]

82 State the color of the indicator bromcresol green if it is added to a sample of the KOH(aq) solution.   [1]

83 Complete the equation on your answer sheet by writing the chemical formula for *each* product.   [1]

84 Show a numerical setup for calculating the molarity of the KOH solution.   [1]

85 Explain, in terms of aqueous ions, why 15.0 mL of a 1.0 M HCl(aq) solution is a better conductor of electricity than 15.0 mL of a 0.010 M HCl(aq) solution.   [1]

Answer Sheet
June 2019

Chemistry—Physical Setting

PART B–2

51

52 _____

_____

_____

53 _____ u

54 _____

55 _____ **K**

56 _____ **g**

57 Boiling point: _____

Freezing point: _____

58 Temperature: _____

Pressure: _____

59 _____ **mL**

60 _____

_____

_____

61 _____

_____

_____

62 _____

_____

_____

63 _____

_____

_____

64 $^{60}_{27}\text{Co} \rightarrow$ _____ $+ \, ^{0}_{-1}\text{e} + \, ^{0}_{0}\gamma$

65 _____ y

PART C

66 _____

67

68 _____ **g**

69 _____ $KClO_3(s) \rightarrow$ _____ $KCl(s) +$ _____ $O_2(g)$

70 _____

71 _____

72

73 _____

_____

_____

74 _____

75 _____

76 _____

77 _____

78 _____

_____

_____

79 _____

80 _____

_____

_____

81 _____

82 _____

83 $HCl(aq) + KOH(aq) \rightarrow$ _____ + _____

84

85

Answers
June 2019

Chemistry—Physical Setting

Answer Key

PART A

1. 3	**7.** 1	**13.** 2	**19.** 3	**25.** 1
2. 3	**8.** 4	**14.** 3	**20.** 3	**26.** 3
3. 4	**9.** 1	**15.** 4	**21.** 2	**27.** 3
4. 1	**10.** 3	**16.** 1	**22.** 4	**28.** 4
5. 4	**11.** 2	**17.** 1	**23.** 2	**29.** 4
6. 3	**12.** 3	**18.** 2	**24.** 2	**30.** 3

PART B–1

31. 2	**35.** 1	**39.** 1	**43.** 4	**47.** 3
32. 4	**36.** 3	**40.** 1	**44.** 4	**48.** 2
33. 3	**37.** 3	**41.** 2	**45.** 3	**49.** 2
34. 4	**38.** 4	**42.** 4	**46.** 3	**50.** 1

PART B–2 and **PART C**. *See* **Answers Explained**.

Answers Explained

PART A

1. **3** Every atom of every element consists of a nucleus surrounded by an electron cloud. The nucleus contains positively charged protons as well as neutrons, which have no charge.

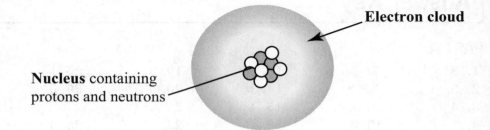

Nucleus containing protons and neutrons

Electron cloud

2. **3** Neutrons are electrically neutral and would not be attracted to either of the electrically charged plates.

WRONG CHOICES EXPLAINED:

(1) Protons are positively charged and would be attracted to the negatively charged plate, since opposite charges attract each other.

(2) Electrons are negatively charged and would be attracted to the positively charged plate, since opposite charges attract each other.

(4) Positrons are positively charged and would be attracted to the negatively charged plate, since opposite charges attract each other.

3. **4** Neutrons and protons have a mass of approximately 1 amu (atomic mass unit), while the mass of an electron is approximately 0.00054 amu. Electrons are so small compared to protons and neutrons that electron mass is ignored when calculating the atomic mass of an atom.

4. **1** Electron shells closer to the nucleus are lower in energy than electron shells farther from the nucleus. This is because negatively charged electrons are attracted to the positively charged nucleus. An electron in the first electron shell of any atom will have less energy than an electron in the second shell.

5. **4** The outermost electrons in an atom are called the valence electrons. Look at the electron configurations of these elements on the Periodic Table in your Chemistry Reference Tables. The last number in each element's electron configuration represents its valence (outermost) electrons. Fluorine, with an electron configuration of 2-7, has 7 valence electrons.

WRONG CHOICES EXPLAINED:
  (1) The electron configuration for sodium (Na) is 2-8-1, meaning that sodium has 1 valence electron.
  (2) The electron configuration for phosphorus (P) is 2-8-5, meaning that phosphorus has 5 valence electrons.
  (3) The electron configuration for nitrogen (N) is 2-5, meaning that nitrogen has 5 valence electrons.

6. **3** Elements differ in their chemical reactivity because of the number and positioning of their valence electrons. Sodium is an alkali metal and is highly reactive in water, whereas silver is chemically unreactive with pure water.

WRONG CHOICES EXPLAINED:
  (1) Mass cannot be used to distinguish between silver and sodium because samples of these metals could have the same mass.
  (2) Volume cannot be used to distinguish between silver and sodium because samples of these metals could have the same volume.
  (4) Both silver and sodium are solid at room temperature.

7. **1** Graphite and diamond are both allotropes of the element carbon. Allotropes are different structural forms of the same element. For example, graphite has a layered structure, and diamonds have a three-dimensional network structure. Because of their different structures, graphite and diamond will have different properties, such as hardness, conductivity, and melting point.

8. **4** Electronegativity and atomic radii values vary from element to element moving both down a group or across a period on the Periodic Table. Refer to Reference Table S to look at these values for C, Si, Ge, Sn, and Pb, and you can confirm that both the atomic radii and electronegativity are different for each element.

WRONG CHOICES EXPLAINED:
  (1), (3) Group 14 elements all have four valence electrons. Find the electron configurations for these elements, which are printed on the Periodic Table. The last number in each element's electron configuration is "4," which represents the valence (outermost) electrons.

(2) Only two electrons are needed to fill the first electron shell in any atom. Looking at the electron configurations on the Periodic Table for Group 14 elements, "2" represents the first shell electrons.

9. **1** Compounds are substances formed from two or more different elements that are chemically bonded in fixed proportions. Remember this definition!

WRONG CHOICES EXPLAINED:
(2) Sodium (Na) is a metal found in Group 1 on the Periodic Table.
(3) Sodium bromide (NaBr) is a solid at room temperature and standard pressure, but this is a physical property and is not sufficient evidence to classify NaBr as a chemical compound. Many substances that are not compounds are solids at the conditions stated in this question, such as gold (an element) and sand (a mixture).
(4) The ability to dissolve is a physical property and again is not sufficient evidence to classify a substance as a chemical compound. For example, a mixture of salt and sugar would dissolve, but in a mixture, the components are physically, not chemically, combined.

10. **3** Chemical formulas provide information about the types and proportion of elements in a compound. The empirical formula provides the smallest whole number ratio of elements in a compound, and a structural formula shows how the atoms are arranged in the compound. Consider the example of 2-pentene shown below:

Molecular Formula   Empirical Formula   Structural Formula
C_5H_{10} CH_2

$$H-\overset{\overset{H}{|}}{C}-C=C-\overset{\overset{H}{|}}{\underset{\underset{H}{|}}{C}}-\overset{\overset{H}{|}}{\underset{\underset{H}{|}}{C}}-H$$

11. **2** Neither charge nor mass nor energy can be created or destroyed during chemical reactions. The charge and mass of the products in a chemical reaction will always equal the charge and mass of the reactants. Energy may change form during a chemical reaction, but the net sum of energy contained in a chemical system and the environment surrounding the system before and after the reaction is the same.

12. **3** Polarity of a covalent bond is caused by the unequal sharing of the electrons in the bond. Electronegativity measures an atom's attraction for a bonded pair of electrons. When atoms with different electronegativities form a covalent bond, the atom with higher electronegativity will have a stronger attraction for the electrons in the bond and will assume a partial negative charge (δ^-), while the atom with lower electronegativity will assume a partial positive charge (δ^+). A larger difference in electronegativity between the atoms in the bond causes stronger partial charges to form, leading to a bond with greater polarity.

13. **2** Magnesium (Mg) is an element found in Group 2 on the Periodic Table. All elements are composed of atoms, and atoms cannot be further broken down during the course of a chemical reaction.

WRONG CHOICES EXPLAINED:
(1), (3), and (4) Ammonia (NH_3), methane (CH_4), and water (H_2O) are all compounds. Compounds are composed of molecules, and molecules can be broken down into their constituent atoms during chemical reactions.

14. **3** Mixtures contain two or more substances (elements or compounds) that are physically mixed together but are not chemically bonded together. The proportions of components in a mixture can vary. For example, consider a solid mixture of salt and sugar. This mixture can contain more salt or less salt, but it is still a mixture of salt and sugar.

WRONG CHOICES EXPLAINED:
(1), (2) Since components of a mixture are not chemically combined, each component retains all of its original properties after mixing.
(4) Proportions of components in mixtures are always variable.

15. **4** Sugar and sand can be separated on the basis of having different solubilities in water. Sugar will dissolve in water, while sand will not dissolve in water. Adding water to a sugar/sand mixture will dissolve the sugar. Passing the water/sugar/sand mixture through a filter will cause sand particles to be trapped by the filter paper, while the dissolved sugar and the water will pass through the filter paper.

16. **1** Gases consist of tiny particles (atoms or molecules) that are very far apart from each other compared with their size. Because their size is so small, the kinetic molecular theory for an ideal gas assumes that the volume of the gas particles themselves is insignificant compared with the volume of the space that they occupy.

WRONG CHOICES EXPLAINED:

(2) Intermolecular forces between gas particles are very weak! This is why gas particles move randomly and separately from each other.

(3) This answer choice describes the spatial arrangement of particles in a solid, not a gas.

(4) This answer choice describes particle arrangement in solid and liquids, not gases. We can see samples of solid and liquid matter because the particles are very close together. Gases are invisible because the tiny particles are spread very far apart.

17. **1** Exothermic phase changes are phase changes in which energy is released. In the process of freezing, the heat of fusion is released, and particles lose potential energy as they assume the closely packed, crystalline arrangement of a solid. In the process of condensation, the transition from the gas to the liquid state, the heat of vaporization is released, and particles lose potential energy as intermolecular forces draw gas particles back together into the much denser liquid phase.

18. **2** Reference Table I shows the heat of reaction (ΔH) for several dissolving equations near the bottom of the table. The heat of reaction for dissolving NaOH is –44.51 kJ. The negative sign indicates that this is an exothermic process in which heat is released, as stated in the footnote beginning with an asterisk (*) under the table.

19. **3** The physical state of a sample of matter is determined by the spacing and arrangement of particles within the sample. As shown in the diagrams below, gases contain particles that are very widely spaced and randomly arranged, whereas solids contain particles that are tightly packed and arranged in a regular, repeating geometric pattern.

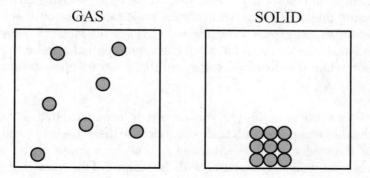

GAS   SOLID

20. **3** A process reaches equilibrium when opposing changes occur at equal rates. An example of a physical change reaching equilibrium is the point at which a solution becomes saturated and the rates of dissolving and precipitation occur at equal rates within the solution. Chemical equilibrium is reached when the rate of the forward reaction equals the rate of the reverse reaction in a reversible reaction. Nuclear changes are not reversible and cannot reach equilibrium.

21. **2** This is a "definition question." The heat of reaction, ΔH, is defined as the potential energy of the products minus the potential energy of the reactants. ΔH can be found on a potential energy diagram, as shown below.

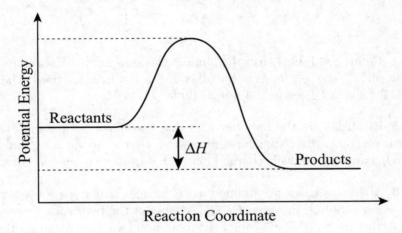

22. **4** Adding a catalyst to a chemical reaction lowers the activation energy (E_a) for the reaction, making it possible for more collisions to have sufficient energy for particles to react.

WRONG CHOICES EXPLAINED:
(1), (2), and (3) None of these are affected by the addition of a catalyst.

23. **2** Systems naturally tend to undergo changes that produce lower total energy and greater disorder or randomness (entropy). For example, when a slab of rock falls from the top of a cliff, it falls to the bottom of the cliff (a position of lower potential energy) and smashes into many pieces (greater randomness).

24. **2** Carbon is a versatile element that forms four bonds to attain a stable octet of eight valence electrons. Because carbon forms four bonds, it is capable of forming many different structures, including chains, rings, and networks, as shown in the diagrams below.

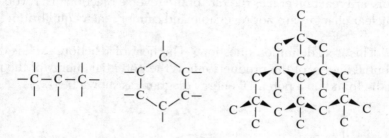

25. **1** Reference Table Q lists the homologous series of hydrocarbons. Ethene, with the suffix "-ene," belongs to the alkene series of hydrocarbons. The general formula for alkenes given in Reference Table Q is C_nH_{2n}.

26. **3** Electrons are the only particles that can move between atoms during chemical reactions. Electrons have a negative charge, so for a neutral fluorine atom to become a negatively charged ion (F^-), it must gain one electron.

27. **3** Arrhenius bases are defined as substances that release hydroxide ions (OH^-) when dissolved in water. $Ca(OH)_2$ contains the hydroxide ion, so it must be the Arrhenius base. Don't forget that common bases are listed on Reference Table L. Arrhenius bases will always contain the OH^- ion.

WRONG CHOICES EXPLAINED:
 (1) HNO_3 is nitric acid, as listed on Reference Table K.
 (2) H_2SO_3 is sulfurous acid, as listed on Reference Table K.
 (4) CH_3COOH is ethanoic acid, as listed on Reference Table K.

28. **4** Nuclear fusion is a nuclear reaction in which two small nuclei fuse together to form a heavier nucleus. This reaction is accompanied by the release of a terrific amount of energy. Hydrogen fusion, for example, is the reaction that powers the sun and the stars.

WRONG CHOICES EXPLAINED:
 (1) Positron emission is a type of natural transmutation reaction.
 (2) Gamma emission occurs during several types of nuclear reactions, including fusion and fission, and often during alpha decay.
 (3) Nuclear fission is the opposite of nuclear fusion. In nuclear fission, a very large nucleus (often U-235) splits into two or more medium-sized nuclei.

29. **4** Nuclear reactions, especially fusion and fission, are dangerous reactions! Consider the destruction caused by the fission bombs that ended World War II. The energy released during nuclear reactions is many orders of magnitude larger than even the most exothermic chemical reactions.

30. **3** U-238 is a radioactive isotope that undergoes radioactive decay in a series of steps, eventually resulting in the formation of Pb-206, a stable isotope of lead. Since the rate of uranium decay into lead is known, finding the ratio of lead to uranium in a rock sample can be used to determine the age of the sample.

PART B–1

31. **2** All elements produce unique bright-line spectra when excited electrons in these atoms fall back to ground state. The bright-line spectrum of a mixture of two or more elements will contain all of the wavelength bands of each separate element superimposed together on the same spectrum. Element *J* can quickly be eliminated since the two wavelength bands at ~730 nm are not evident in the mixture. If element *J* is not in the mixture, the answer must be choice (2), and careful inspection shows that all wavelength bands from elements *G* and *L* are also found in the mixture.

32. **4** The Periodic Table in the Chemistry Reference Tables shows the ground state configuration of chlorine as 2-8-7. A chlorine atom would be in an excited state if its configuration were 2-7-8 because this configuration shows that a second shell electron has moved to the third shell.

WRONG CHOICES EXPLAINED:
(1) Chlorine atoms have 17 electrons. The electron configuration 2-8-7-2 has a total of 19 electrons and is an excited state electron configuration for a potassium atom.
(2) This is the ground state electron configuration for a chlorine atom. The question asks for an *excited* state configuration.
(3) 2-8-8 is the ground state electron configuration for an atom of argon.

33. **3** The actual value for the density of aluminum, 2.70 g/cm^3, is found on Reference Table *S*. The calculated value of the density of aluminum, as stated in the question, is 2.85 g/cm^3. Use the percent error formula, found on Reference Table *T*, to set up the calculation for percent error:

$$\% \, error = \frac{\text{measured value} - \text{accepted value}}{\text{accepted value}} \times 100$$

Plug in the values from the question:

$$\% \text{ error} = \frac{2.85\,\text{g/cm}^3 - 2.70\,\text{g/cm}^3}{2.70\,\text{g/cm}^3} \times 100$$

$$\% \text{ error} \cong 5.6\%$$

34. **4** Find the elements magnesium and calcium on the Periodic Table in your Reference Tables. Looking at the electron configurations for both elements, we see that each element has 2 electrons in the outermost shell. Electrons in the outermost shell are known as valence electrons. Valence electrons are involved in forming chemical bonds between atoms and are therefore largely responsible for an atom's chemical properties.

WRONG CHOICES EXPLAINED:
(1) All elements, no matter how different their properties are, have atoms with equal numbers of protons and electrons, so this cannot be responsible for the chemical similarity between magnesium and calcium.
(2) Calcium atoms contain 20 protons, whereas magnesium atoms contain 12 protons.
(3) All atoms with two or more electrons in the ground state have two electrons in the first shell. The first shell has the least energy of all electrons shells since it is closest to the nucleus, and electrons in the ground state will occupy the lowest energy positions possible.

35. **1** The atomic radius decreases as elements are considered from left to right across Period 2 of the Periodic Table. This trend is caused by increasing nuclear charge, which draws the electron cloud in closer, thereby shrinking the atomic radius. Since atomic radius data is provided in Reference Table S, you can always look up values and figure out the trend if you forget.

WRONG CHOICES EXPLAINED:
(2) Electronegativity values increase as elements are considered from left to right across Period 2 of the Periodic Table because increasing nuclear charge increases an element's attraction for electrons.
(3) Ionization energy values increase as elements are considered from left to right across Period 2 of the Periodic Table because increasing nuclear charge makes it more difficult to remove electrons from an atom.
(4) The number of protons increases as elements are considered from left to right across Period 2 of the Periodic Table, and nuclear charge is caused by the charge of the protons in the nucleus.

36. **3** All coefficients in chemical reactions represent particles or moles. Mole ratios can be used to find the number of moles of any substance in a chemical equation if the moles of another substance are known. Dimensional analysis or mole ratios can be used to solve this problem:

Using dimensional analysis:

$$\frac{5.0 \;\text{mol} \;\text{C}_4\text{H}_{10}}{1} \times \frac{8 \,\text{mol}\, \text{CO}_2}{2 \;\text{mol} \;\text{C}_4\text{H}_{10}} = 20. \,\text{mol}\, \text{CO}_2$$

Using mole ratios:

$$\frac{5.0 \,\text{mol}\, \text{C}_4\text{H}_{10}}{2} = \frac{x \,\text{mol}\, \text{CO}_2}{8}$$

$$x = 20. \,\text{mol}\, \text{CO}_2$$

37. **3** Use the percent composition formula found on Reference Table *T*:

$$\% \text{ composition by mass} = \frac{\text{mass of part}}{\text{mass of whole}} \times 100$$

$$\% \text{ composition by mass} = \frac{(14.0 \,\text{g/mol})(2)}{32.0 \,\text{g/mol}} \times 100$$

$$\% \text{ composition by mass} \cong 88\%$$

38. **4** The oxygen atom has 8 electrons, but the atom must gain two electrons to form the O^{2-} ion. The ion has 10 electrons, just like all neon atoms.

WRONG CHOICES EXPLAINED:
(1) Ca^{2+} ions have 18 electrons and are isoelectronic with argon.
(2) Cl^- ions have 18 electrons and are isoelectronic with argon.
(3) Li^+ ions have just two electrons and are isoelectronic with helium.

39. **1** Higher boiling points are caused by stronger intermolecular forces between molecules in a liquid sample. Boiling involves a transition from the liquid state to the gas state; as boiling occurs, molecules gain the potential energy that they need to overcome intermolecular attractions and separate from each other to form a gas. When intermolecular forces are strong, substances will heat to a higher temperature in the liquid state before boiling occurs since it is harder for particles to overcome the intermolecular attractions and form a gas.

40. **1** Xenon exists as a monatomic gas at STP.

WRONG CHOICES EXPLAINED:
(2) Xenon is not a diatomic gas.
(3) Xenon is not in the solid phase at STP.
(4) Xenon is not in the liquid phase at STP.

41. **2** Heat formulas are found on Reference Table T. Since the temperature of the water in the beaker increases from 20.°C to 35°C, the correct formula is:

$$q = mC\Delta T$$

where $q =$ the amount of heat in joules (J) lost or gained, $m =$ mass of the water in grams, $C =$ specific heat capacity of water, and $\Delta T =$ change in temperature.

In this case, $\Delta T = T_{final} - T_{initial} = 35 - 20. = 15°C$. Plugging in the given values leads to the following expression:

$$q = (75 \text{ g})(4.18 \text{ J/g·K}) (15 \text{ C}) \cong 4700 \text{ J}$$

(**NOTE:** The specific heat capacity of water (C) is listed on Reference Table B.)

42. **4** Anyone wishing to increase the rate of a chemical reaction needs to increase the number of effective collisions between reactant particles. Increasing the temperature and increasing the concentration of the HCl solution will increase the total number of collisions between reactant particles, and the more total collisions there are, the more effective collisions there will be.

43. **4** The reaction is at equilibrium, so the rate of the forward reaction must equal the rate of the reverse reaction. Since both products and reactants are forming at equal rates, the concentration of each remains constant.

44. **4** Systems at equilibrium prefer to remain at equilibrium. Since this is an endothermic reaction (the heat term is on the reactant side of the equation), heat is absorbed in the forward direction. If this system is stressed by increasing the temperature, the system will "shift to the right" (temporarily favor the forward direction) so that some of the excess heat will be absorbed, which will lower the temperature and restore equilibrium.

WRONG CHOICES EXPLAINED:

(1) Adding a catalyst will cause the reaction rate to increase in *both* the forward and the reverse direction, but equilibrium will not be affected.

(2) Adding more $PCl_3(g)$ will cause a shift to the left, since the reaction will try to use up the excess $PCl_3(g)$ by reacting in the reverse direction.

(3) Increasing the pressure on this system will cause a shift to the left, since there are fewer moles of gas on the reactant side.

45. **3** This organic compound is classified as an amine since the "$-NH_2$" functional group is present (see Reference Table R). The suffix for amines is "–amine."

WRONG CHOICES EXPLAINED:

(1) Pentanone is a ketone.

(2) 1-pentanol is an alcohol.

(4) Pentanamide is an amide. Although amides also contain nitrogen, the compound in this question does not contain oxygen and therefore cannot be an amide.

46. **3** This organic compound is a hydrocarbon, since it contains only carbon and hydrogen. Reference Table Q shows that alkynes contain triple bonds between carbon atoms.

47. **3** Fermentation reactions involve the breakdown of glucose into ethanol and carbon dioxide in the presence of an enzyme catalyst.

WRONG CHOICES EXPLAINED:

(1) Although this reaction forms ethanol, glucose is not a reactant. This is a synthesis reaction.

(2) This is a type of organic reaction known as an addition reaction.

(4) This reaction forms an aldehyde, not ethanol.

(NOTE: You need to be able to recognize seven major types of organic reactions. These include addition, substitution, esterification, polymerization, fermentation, saponification, and combustion. Memorize these reactions!)

48. **2** As a reactant in the compound $CuCl_2$, copper has an oxidation number of $+2$. As a pure element on the product side, copper has an oxidation number of 0.

49. **2** The Brønsted-Lowry theory of acids and bases defines an acid as a proton (H^+) donor and a base as a proton (H^+) acceptor. $CH_3COOH(aq)$ must lose (donate) an H^+ in the forward reaction in order to form the product $CH_3COO^-(aq)$. Likewise, in the reverse reaction, $H_3O^+(aq)$ must lose an H^+ to form $H_2O(\ell)$.

50. **1** Natural transmutation, also known as radioactive decay, involves the spontaneous disintegration of an unstable nucleus. This type of nuclear reaction can always be recognized by the presence of only one substance on the reactant side and the presence of a decay particle (in this case, an alpha particle, ^4_2He) on the product side.

WRONG CHOICES EXPLAINED:

(2) Artificial transmutation reactions take place when a high-speed particle (in this case an alpha particle, ^4_2He) collides with a target nucleus.

(3) Fission reactions occur when a high-speed neutron collides with a U-235 nucleus, forcing it to split into two or more smaller nuclei. The formation of neutrons enables a chain reaction to occur.

(4) Nuclear fusion happens when two lightweight nuclei merge to form heavier nuclei.

(NOTE: You need to be able to recognize the four types of nuclear reactions. These include natural transmutation, artificial transmutation, fission, and fusion. Memorize these reactions!)

PART B–2

[All questions in Part B–2 are worth 1 point.]

51. The suffix "-al" in the compound methanal reveals that this compound is an aldehyde (see Reference Table R). The general formula for compounds that contain the aldehyde functional group is shown below:

$$
\begin{array}{c}
\text{O} \\
\parallel \\
\text{C} \\
\diagup \quad \diagdown \\
\text{R} \qquad \text{H}
\end{array}
$$

Specifically, however, the compound methanal contains only one carbon atom (refer to Reference Table P), so the structural formula of methanal is

52. All atoms are electrically neutral because the **number of positively charged protons equals the number of negatively charged electrons**.

(NOTE: This question requires the response to be "in terms of subatomic particles." Acceptable responses must either use the term "subatomic particles" directly or must use the specific names of relevant subatomic particles [protons and electrons].)

53. The formula mass, also known as the molar mass, is calculated by adding the atomic masses of the individual atoms in a compound. The information that precedes the question states that the formula of heavy water is D_2O. The table includes atomic masses for three isotopes of the element hydrogen. Heavy water is composed of two atoms of deuterium (H-2) and one atom of oxygen. Oxygen's atomic mass is found on the Periodic Table in the Reference Tables.

Atomic mass of D_2O = 2(atomic mass of H-2) + 1 (atomic mass of oxygen)

Atomic mass of D_2O = 2(2.0141 u) + 16.0 u

Atomic mass of D_2O = **20.03 u**

(NOTE: The abbreviation "u" stands for "atomic mass unit," or "amu." Acceptable answers included all values from **19.999 u to 20.03 u**, inclusive. Significant figures were not considered when grading this question.)

54. Reference Table N contains information about the decay mode for a number of radioactive isotopes. Consult Reference Table N to find the decay mode of tritium, H-3. Tritium undergoes **beta decay**.
(NOTE: Other acceptable answers include **beta particle, beta, β^-, $_{-1}^{0}e$**, or $_{-1}^{0}\beta$.)

55. The formula to convert temperatures from degrees Celsius to the Kelvin scale is found on the last page of your Reference Tables, Reference Table T:

$$K = °C + 273$$

Plug in the given Celsius temperature to solve for the temperature on the Kelvin scale:

$$K = 23°C + 273 = \textbf{296 K}$$

56. Reference Table G shows the solubilities of different solutes in water over a range of temperatures. The lines on the graph represent lines of saturation, meaning that the solution is holding the maximum amount of solute possible.

At 23°C, Reference Table G shows that approximately 90.0 g of $NaNO_3$ will dissolve in 100.0 g of water. The amount of $NaNO_3$ in the initial solution must be subtracted from this calculation to find the additional amount of $NaNO_3$ that would need to be added to reach saturation:

$$90.0 \text{ g} - 85.0 \text{ g} = \textbf{5.0 g of } \mathbf{NaNO_3} \textbf{ must be added}$$

(NOTE: Credit was awarded for any value from 4.0 g to 6.0 g, inclusive.)

57. Boiling point elevation and freezing point depression are two colligative properties of solutions. Dissolved particles raise the boiling point and lower the freezing point of a solution. The more concentrated the solution, the greater the effect on the boiling point and the freezing point. This question is tricky because, instead of adding more solute particles to the solution, *water* is added, leading to a *less* concentrated solution. Having a solution that is less concentrated would result in **a lower boiling point and a higher freezing point**.

58. Real gases behave more like ideal gases when temperature is high and pressure is low. These conditions minimize any possible attractions between gas particles, allowing the gas to adhere more closely to the principles of the Kinetic Molecular Theory of Gases. The original conditions for the temperature and pressure of the gas sample, as described in the information that precedes the question, are 296 K and 101.3 kPa, respectively. This sample will behave more like an ideal gas if **the temperature was higher and the pressure was lower**.
(NOTE: Acceptable responses also included "the temperature increased and the pressure decreased" or "any temperature above 296 K and any pressure lower than 101.3 kPa.")

59. Use the Combined Gas Law Equation on Reference Table T to solve this problem:

$$\frac{P_1 V_1}{T_1} = \frac{P_2 V_2}{T_2}$$

Substitute information from the question:

$$\frac{(101.3 \, \text{kPa})(200. \, \text{mL})}{296 \, \text{K}} = \frac{(152.0 \, \text{kPa})(V_2)}{336 \, \text{K}}$$

Rearrange the equation:

$$V_2 = \frac{(101.3\,\text{kPa})(200.\,\text{mL})(336\,\text{K})}{(296\,\text{K})(152.0\,\text{kPa})}$$

Solve this equation to find that **V_2 = 151.3 mL**.

(NOTE: Credit was awarded for any value from 151 mL to 151.4 mL, inclusive.)

60. Temperature is directly related to the kinetic energy of particles in a sample of matter. Kinetic energy is the energy of particle motion. Putting these two ideas together leads to the understanding that particles in a sample of matter move faster at higher temperatures and slower at lower temperatures. Picture gas particles in an enclosed space (see the image below). Imagine the gas particles moving slower at a lower temperature. **They would collide with each other less often, and they would also collide with less force.** A good analogy would be bumper cars in a bumper car arena. At lower speeds, the cars collide less often and with considerably less force than they do at higher speeds.

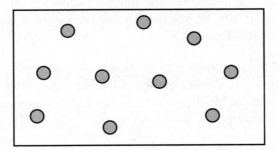

61. The introductory information for this question states that the sample of $CO_2(g)$ is in a *sealed*, rigid container. If the container is sealed, then gas particles cannot enter or leave the container. If the number of gas particles stays constant, then **the mass of gas in the container also stays constant**!

62. Isotopes are different versions of the same element that have different numbers of neutrons. Co-59 and Co-60 are isotopes for two reasons: (1) they have the **same number of protons** (27, the atomic number of cobalt) and (2) they have **different numbers of neutrons**. Co-60 has a mass number of 60. To find the number of neutrons, subtract the number of protons from the mass number: $60 - 27 = 33$ neutrons. Using the same calculation, the number of neutrons in Co-59 is $59 - 27 = 32$ neutrons.

63. **Gamma radiation has greater penetrating power than beta emission.** Gamma radiation is pure energy and has neither mass nor charge. Just as X-rays can penetrate the body to expose the location of a broken bone, gamma radiation can pass right through matter. Beta radiation, on the other hand, has a very small mass and a negative charge. Having a charge causes the beta particles to slow down as they are repelled by electrons and are attracted to nuclei in the matter through which they are traveling.

64. Like regular chemical reactions, nuclear reactions must be balanced for both mass and charge. To balance a nuclear equation, the sum of the atomic numbers (bottom) and the mass numbers (top) must be equal on both sides of the arrow. Consider the reaction below:

$$^{60}_{27}\text{Co} \rightarrow \underline{\hspace{2cm}} + ^{0}_{-1}\text{e} + ^{0}_{0}\gamma$$

Since the beta particle and the gamma radiation both have a mass number of 0, the missing product must have a mass number of 60. To solve for the atomic number of the missing product, ask yourself, "What number, when added to −1, will equal 27?" The atomic number of the missing product must be 28. Look on the Periodic Table to find that the missing product is an atom of **nickel-60**.

(NOTE: $^{60}_{28}$**Ni**, **Ni-60**, and ^{60}**Ni** are all acceptable responses as well.)

65. One half-life is the time required for a radioisotope to decay to half of its original mass. To find the total number of half-lives required for Co-60 to decay from an original mass of 80.00 grams to a mass of 10.00 grams, divide the mass in half over a series of steps.

In the first half-life, Co-60 decays from 80.00 grams to 40.00 grams. (The other 40.00 grams undergo radioactive decay.)

In the second half-life, the 40.00 grams of Co-60 that remain decay to 20.00 grams.

In the third half-life, the 20.00 grams of Co-60 that remain decay to 10.00 grams.

Three half-lives are needed for the radioactive decay of Co-60 from 80.00 grams to 10.00 grams. Reference Table N shows that Co-60 has a half-life of 5.271 years. The total time required for three half-lives is 3×5.271 years = **15.813 years**.

PART C

[All questions in Part C are worth 1 point.]

66. $KClO_3$ is an ionic compound that is formed when the K^+ ion bonds to the ClO_3^- ion. To name this compound, look up the name of ClO_3^- in Reference Table E: chlorate. Put this together with the name of the cation to find the compound name: **potassium chlorate**.

67. The data table shows that 0.80 g of $O_2(g)$ are produced. To convert this mass into moles of $O_2(g)$, use the following formula found on Reference Table T:

$$\text{number of moles} = \frac{\text{given mass}}{\text{gram-formula mass}}$$

Plug in the numbers:

$$\textbf{number of moles} = \frac{\textbf{0.80 g O}_2}{\textbf{32 g/mol O}_2}$$

(NOTE: This question asks ONLY for the numerical setup to determine the number of moles of $O_2(g)$ produced. The numerical setup is all that you should record on your answer sheet. If you would like to check your work by solving for the number of moles, do this on scrap paper.)

68. While stoichiometry can be used to solve this problem, there is a much easier and faster method, using the Law of Conservation of Mass. The unbalanced equation is:

$$KClO_3(s) \rightarrow KCl(s) + O_2(g)$$

We know the starting mass of $KClO_3(s)$ (2.07 g), and we know the mass of $O_2(g)$ produced (0.80 g). Since matter is neither created nor destroyed in a chemical reaction, the total mass of the products must equal the mass of the original reactant. Subtract the mass of $O_2(g)$ produced from the mass of $KClO_3(s)$ used to find the mass of $KCl(s)$ produced:

$$2.07 \text{ g} - 0.80 \text{ g} = \textbf{1.27 g of KCl(s)}$$

69. To balance a chemical equation, the number of each type of atom must be equal on both sides of the equation. The unbalanced equation is shown below:

$$\underline{\quad} \; KClO_3(s) \rightarrow \underline{\quad} \; KCl(s) + \underline{\quad} \; O_2(g)$$

As written, there is 1 potassium atom and 1 chlorine atom on each side of the equation, so these two elements are already balanced. However, there are 3 oxygen atoms on the left of the reaction arrow and only 2 oxygen atoms on the right side of the equation, so the equation as it stands is not balanced for oxygen. This odd/even situation often occurs when balancing equations that contain diatomic elements, like O_2. The best way to handle this situation is to *double all of the coefficients* (except the coefficient for the O_2, which isn't known yet). Doubling an odd number results in an even number. The finished balanced equation becomes:

$$\underline{\;2\;} \; KClO_3(s) \rightarrow \underline{\;2\;} \; KCl(s) + \underline{\;3\;} \; O_2(g)$$

70. If you have any doubt about which of the ions listed are the hydrogen carbonate ions, consult Reference Table *E*. The table for this question shows that 0.180 g of HCO_3^- ions are present in a 500. g sample of bottled water. The number 0.180 contains **three significant figures**. The zero in the thousandths place does not change the absolute value of the number, but it does show the precision of the measurement.

71. The formula for the sulfate ion, SO_4^{2-}, can be found in Reference Table *E*. Reference Table *F* contains information about the solubility of different ions in aqueous solutions. SO_4^{2-} is listed as soluble under most conditions *except* when combined with the five cations listed in the second column of that table. If any of these five ions is present, an insoluble compound will form. Of the ions listed in the table for this question, only **Ca^{2+}** will form an insoluble compound in the presence of SO_4^{2-} in the bottled water.

72. Look at Reference Table *T* to find the formula for parts per million:

$$\text{parts per million} = \frac{\text{mass of solute}}{\text{mass of solution}} \times 1{,}000{,}000$$

Next, plug in the information for this question:

$$\textbf{parts per million} = \frac{\textbf{0.0033 g}}{\textbf{500. g}} \times \textbf{1,000,000}$$

(NOTE: This question asks ONLY for the numerical setup for the calculation of the concentration of Na^+ ions in parts per million. The numerical setup is all that you should record on your answer sheet. If you would like to check your work by solving for the number of ppm, do this on scrap paper.)

73. An Mg^{2+} ion must gain two electrons to become a neutral Mg atom. The Mg^{2+} ion has an electron configuration of 2-8. The Mg atom has an electron configuration of 2-8-2. **The Mg atom has a larger radius than that of the Mg^{2+} ion** since the Mg atom has electrons in the third electron shell, whereas the Mg^{2+} ion has no electrons in the third electron shell.

74. Organic chemistry, by definition, is the chemistry of **carbon**-based compounds. Carbon atoms form four covalent bonds and can easily bond with other carbon atoms to form chains, rings, and networks.

75. An empirical formula shows the elements of a compound in their lowest whole number ratio. Use the structural diagram given for this question, and count the atoms to find the molecular formula, $C_4H_8O_2$. Next, reduce this molecular formula to lowest terms by dividing each subscript by the common factor of 2. Thus, the empirical formula is **C_2H_4O**.
(NOTE: The order of elements in the answer may vary.)

76. The functional group for the organic compound in this question is marked in the diagram below:

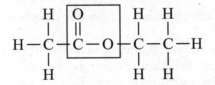

Look at the second column in Reference Table R, which shows functional groups for several classes of organic compounds. This compound is an **ester**.

77. All of the covalent bonds between hydrogen atoms and carbon atoms in the molecule are single covalent bonds, represented by C–H. There are **two electrons shared** in every single covalent bond.

78. The salt bridge **allows ions to flow into each compartment in the voltaic cell in order to maintain electrical neutrality** in the system. A buildup of positive or negative charges in either compartment would cause electrons to stop flowing.

79. Magnesium is higher than silver on Reference Table *J*, so magnesium will *oxidize* in the presence of silver. Oxidation is defined as the "loss of electrons." Magnesium will lose two electrons to form Mg^{2+} ions as shown in the following half-cell equation:

$$Mg(s) \rightarrow Mg^{2+}(aq) + 2e^-$$

80. **Voltaic cells produce electrical energy, while electrolytic cells use electrical energy.** Voltaic cells are commonly known as batteries. In batteries, a spontaneous redox reaction generates electricity, whereas in electrolytic cells, electricity from an external power source is used to drive a nonspontaneous redox reaction.

81. pH measures the concentration of hydrogen ion, H^+, in aqueous solution. pH is a logarithmic function, meaning that it is based on the exponent of the base 10 molarity of the hydrogen ion content. Each pH value is 10 times different in hydrogen ion molarity than the pH immediately higher and lower than itself. Hydrogen ions make a solution acidic, and solutions that are more acidic have lower pH values. The problem states that the original HCl solution had a pH of 2. The pH of a solution that is ten times *less* acidic will be one number *higher* than the original pH—in other words, a pH of **3**.

82. KOH(aq) is a base that is listed on Reference Table *L*. All basic solutions have pH values greater than 7. The KOH solution will be **blue** in the presence of the bromcresol green indicator, since Reference Table *M* shows that bromcresol green is blue in all solutions with a pH greater than 5.4.

83. The reaction between HCl(aq) and KOH(aq) is a neutralization reaction. Neutralization reactions are really double replacement reactions between acids and bases. The word equation for neutralization reactions between Arrhenius acids and bases is:

$$ACID + BASE \rightarrow WATER + SALT$$

Water is always a product of classic neutralization reactions, and the salt that forms completes the double replacement reaction (ions switching places):

$$HCl(aq) + KOH(aq) \rightarrow \textbf{H}_2\textbf{O}(\ell) + \textbf{KCl(aq)}$$

84. Titration is the lab technique that is used to determine the concentration of an acid or a base by neutralizing it with an acid or a base of known concentration. The titration formula is found on Reference Table T:

$$M_A V_A = M_B V_B$$

in which M_A = molarity of H^+, V_A = volume of acid,

M_B = molarity of OH^-, and V_B = volume of base

Substitute the values from this question:

$$(0.010\,M)(15.0\,mL) = (M_B)(7.5\,mL)$$

$$M_B = \frac{(0.010\,M)(15.0\,mL)}{7.5\,mL}$$

(NOTE: Once again, this question asks ONLY for the numerical setup for calculating the molarity of the KOH solution. The numerical setup is all that you should record on your answer sheet. If you would like to check your work by solving for the molarity of the KOH solution, do this on scrap paper.)

85. HCl is a strong electrolyte because it dissociates completely into freely moving ions in an aqueous solution. Freely moving dissociated ions allow charge to flow through the solution. **The 1.0 M HCl solution has a greater concentration of mobile ions**, making it possible for more electrical current to flow through the solution.

Mark (✓) the questions you answered correctly. Count the number of checks and follow the formulas given to determine your score on each topic.

Core Area	□ Questions Answered Correctly

33, 70: (2)

Section M—Math Skills
□ Number of checks ÷ 2 × 100 = ___ %

61: (1)

Section R—Reading Skills
□ Number of checks ÷ 1 × 100 = ___ %

1–5, 31, 32, 52, 62: (9)

Section I—Atomic Concepts
□ Number of checks ÷ 9 × 100 = ___ %

6–8, 34, 35: (5)

Section II—Periodic Table
□ Number of checks ÷ 5 × 100 = ___ %

9–11, 36, 37, 53, 66–69, 75: (11)

Section III—Moles/Stoichiometry
□ Number of checks ÷ 11 × 100 = ___ %

12, 38, 39, 71, 73, 77: (6)

Section IV—Chemical Bonding
□ Number of checks ÷ 6 × 100 = ___ %

13–19, 40, 41, 55–60, 72: (16)

Section V—Physical Behavior of Matter
□ Number of checks ÷ 16 × 100 = ___ %

20–23, 42–44: (7)

Section VI—Kinetics and Equilibrium
□ Number of checks ÷ 7 × 100 = ___ %

Core Area	☐ Questions Answered Correctly

24, 25, 45–47, 51, 76: (7)

Section VII—Organic Chemistry
☐ Number of checks ÷ 7 × 100 = ___ %

26, 48, 74, 78–80: (6)

Section VIII—Oxidation-Reduction
☐ Number of checks ÷ 6 × 100 = ___ %

27, 49, 81–85: (7)

Section IX—Acids, Bases, and Salts
☐ Number of checks ÷ 7 × 100 = ___ %

28–30, 50, 54, 63–65: (8)

Section X—Nuclear Chemistry
☐ Number of checks ÷ 8 × 100 = ___ %

Examination August 2019
Chemistry—Physical Setting

PART A

Answer all questions in this part.

Directions (1–30): For *each* statement or question, write in the answer space the *number* of the word or expression that, of those given, best completes the statement or answers the question. Some questions may require the use of the *2011 Edition Reference Tables for Physical Setting/Chemistry*.

1 Which statement describes the earliest model of the atom?

(1) An atom is an indivisible hard sphere.
(2) An atom has a small, dense nucleus.
(3) Electrons are negative particles in an atom.
(4) Electrons in an atom have wave-like properties.

1 _____

2 In all atoms of bismuth, the number of electrons must equal the

(1) number of protons
(2) number of neutrons
(3) sum of the number of neutrons and protons
(4) difference between the number of neutrons and protons

2 _____

3 Which symbol represents a particle that has a mass approximately equal to the mass of a neutron?

(1) α (3) β^-
(2) β^+ (4) p

3 _____

4 An orbital is a region in an atom where there is a high probability of finding

    (1) an alpha particle   (3) a neutron

    (2) an electron   (4) a positron   4 _____

5 Which electron shell in an atom of calcium in the ground state has an electron with the greatest amount of energy?

    (1) 1   (3) 3

    (2) 2   (4) 4   5 _____

6 As the elements in Period 2 are considered in order from lithium to fluorine, there is an increase in the

    (1) atomic radius

    (2) electronegativity

    (3) number of electron shells

    (4) number of electrons in the first shell   6 _____

7 Which element is classified as a metalloid?

    (1) boron   (3) sulfur

    (2) potassium   (4) xenon   7 _____

8 Strontium and barium have similar chemical properties because atoms of these elements have the same number of

    (1) protons   (3) electron shells

    (2) neutrons   (4) valence electrons   8 _____

9 Which term represents the fixed proportion of elements in a compound?

    (1) atomic mass   (3) chemical formula

    (2) molar mass   (4) density formula   9 _____

10 Which two terms represent types of chemical formulas?

    (1) mechanical and structural

    (2) mechanical and thermal

    (3) molecular and structural

    (4) molecular and thermal   10 _____

11 Which element has metallic bonds at room temperature?

  (1) bromine   (3) krypton

  (2) cesium   (4) sulfur   11 _____

12 What is the number of electrons shared between the atoms in a molecule of nitrogen, N_2?

  (1) 8   (3) 3

  (2) 2   (4) 6   12 _____

13 Given the equation representing a reaction:

$$H + H \rightarrow H_2$$

What occurs during this reaction?

  (1) A bond is broken and energy is absorbed.

  (2) A bond is broken and energy is released.

  (3) A bond is formed and energy is absorbed.

  (4) A bond is formed and energy is released.   13 _____

14 An atom of which element has the strongest attraction for electrons in a chemical bond?

  (1) chlorine   (3) phosphorus

  (2) carbon   (4) sulfur   14 _____

15 At STP, a 50.-gram sample of $H_2O(\ell)$ and a 100.-gram sample of $H_2O(\ell)$ have

  (1) the same chemical properties

  (2) the same volume

  (3) different temperatures

  (4) different empirical formulas   15 _____

16 Which statement describes a mixture of sand and water at room temperature?

  (1) It is heterogeneous, and its components are in the same phase.

  (2) It is heterogeneous, and its components are in different phases.

  (3) It is homogeneous, and its components are in the same phase.

  (4) It is homogeneous, and its components are in different phases.   16 _____

17 Distillation is a process used to separate a mixture of liquids based on different

(1) boiling points   (3) freezing points
(2) densities   (4) solubilities   17 _____

18 According to the kinetic molecular theory, which statement describes the particles in a sample of an ideal gas?

(1) The particles are constantly moving in circular paths.
(2) The particles collide, decreasing the total energy of the system.
(3) The particles have attractive forces between them.
(4) The particles are considered to have negligible volume.   18 _____

19 Which sample of matter has the greatest distance between molecules at STP?

(1) $N_2(g)$ (3) $C_6H_{14}(\ell)$
(2) $NH_3(aq)$ (4) $C_6H_{12}O_6(s)$ 19 _____

20 For a chemical system at equilibrium, the concentrations of both the reactants and the products must

(1) decrease   (3) be constant
(2) increase   (4) be equal   20 _____

21 In terms of disorder and energy, systems in nature have a tendency to undergo changes toward

(1) less disorder and lower energy
(2) less disorder and higher energy
(3) greater disorder and lower energy
(4) greater disorder and higher energy   21 _____

22 The only two elements in alkenes and alkynes are

(1) carbon and nitrogen
(2) carbon and hydrogen
(3) oxygen and nitrogen
(4) oxygen and hydrogen   22 _____

23 Which functional group contains a nitrogen atom and an oxygen atom?

  (1) ester   (3) amide

  (2) ether   (4) amine   23 _____

24 When a sample of Mg(s) reacts completely with $O_2(g)$, the Mg(s) loses 5.0 moles of electrons. How many moles of electrons are gained by the $O_2(g)$?

  (1) 1.0 mol   (3) 5.0 mol

  (2) 2.5 mol   (4) 10.0 mol   24 _____

25 Which statement describes the reactions in an electrochemical cell?

  (1) Oxidation occurs at the anode, and reduction occurs at the cathode.

  (2) Oxidation occurs at the cathode, and reduction occurs at the anode.

  (3) Oxidation and reduction both occur at the cathode.

  (4) Oxidation and reduction both occur at the anode.   25 _____

26 A 0.050 M aqueous solution of which compound is the best conductor of electric current?

  (1) C_3H_7OH (3) $MgSO_4$

  (2) $C_6H_{12}O_6$ (4) K_2SO_4 26 _____

27 What is the color of bromcresol green indicator in a solution with a pH value of 2.0?

  (1) blue   (3) red

  (2) green   (4) yellow   27 _____

28 Which formula can represent hydrogen ions in an aqueous solution?

  (1) $OH^-(aq)$ (3) $H_3O^+(aq)$

  (2) $Hg_2{}^{2+}(aq)$ (4) $NH_4{}^+(aq)$ 28 _____

29 In which reaction is an atom of one element converted into an atom of another element?

(1) combustion
(2) fermentation
(3) oxidation-reduction
(4) transmutation   29 _____

30 In which type of nuclear reaction do nuclei combine to form a nucleus with a greater mass?

(1) alpha decay   (3) fusion
(2) beta decay   (4) fission   30 _____

PART B–1

Answer all questions in this part.

Directions (31–50): For *each* statement or question, write in the answer space the *number* of the word or expression that, of those given, best completes the statement or answers the question. Some questions may require the use of the *2011 Edition Reference Tables for Physical Setting/Chemistry.*

31 The bright-line spectra produced by four elements are represented in the diagram below.

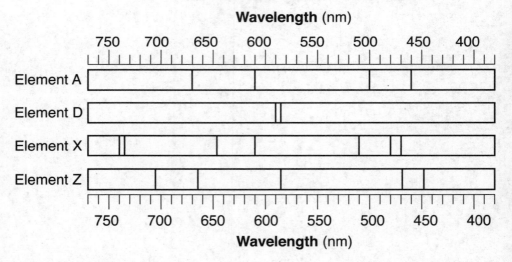

Bright-Line Spectra of Four Elements

Given the bright-line spectrum of a mixture formed from two of these elements:

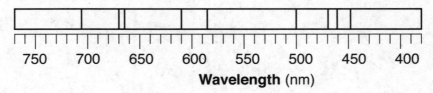

Which elements are present in this mixture?

(1) *A* and *X*   (3) *D* and *X*

(2) *A* and *Z*   (4) *D* and *Z*   31 _____

32 Which electron configuration represents the electrons in an atom of sulfur in an excited state?

(1) $2 - 8 - 6$ (3) $2 - 8 - 7$
(2) $2 - 7 - 7$ (4) $2 - 7 - 8$ 32 _____

33 Which notations represent atoms that have the same number of protons but a different number of neutrons?

(1) H-3 and He-3   (3) Cl-35 and Cl-37
(2) S-32 and S-32   (4) Ga-70 and Ge-73   33 _____

34 What is the chemical name of the compound NH_4SCN?

(1) ammonium thiocyanate
(2) ammonium cyanide
(3) nitrogen hydrogen cyanide
(4) nitrogen hydrogen sulfate   34 _____

35 Which equation represents a conservation of atoms?

(1) $2Fe + 2O_2 \rightarrow Fe_2O_3$
(2) $2Fe + 3O_2 \rightarrow Fe_2O_3$
(3) $4Fe + 2O_2 \rightarrow 2Fe_2O_3$
(4) $4Fe + 3O_2 \rightarrow 2Fe_2O_3$ 35 _____

36 Which compound has covalent bonds?

(1) H_2O (3) Na_2O
(2) Li_2O (4) K_2O 36 _____

37 Which particle diagram represents a sample of oxygen gas at STP?

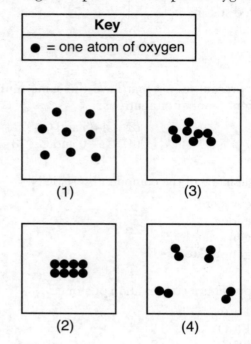

37 _____

38 At which temperature and pressure will a sample of neon gas behave most like an ideal gas?

(1) 300. K and 2.0 atm   (3) 500. K and 2.0 atm
(2) 300. K and 4.0 atm   (4) 500. K and 4.0 atm

38 _____

39 What is the molarity of 2.0 liters of an aqueous solution that contains 0.50 mole of potassium iodide, KI?

(1) 1.0 M   (3) 0.25 M
(2) 2.0 M   (4) 0.50 M

39 _____

40 The volumes of four samples of gaseous compounds at 298 K and 101.3 kPa are shown in the table below.

Sample	Compounds	Volume (L)
1	$NH_3(g)$	44.0
2	$CO_2(g)$	33.0
3	$HF(g)$	44.0
4	$CH_4(g)$	22.0

Which two samples contain the same number of molecules?

(1) 1 and 2   (3) 2 and 3
(2) 1 and 3   (4) 2 and 4   40 _____

41 Hydrochloric acid reacts faster with powdered zinc than with an equal mass of zinc strips because the greater surface area of the powdered zinc

(1) decreases the frequency of particle collisions
(2) decreases the activation energy of the reaction
(3) increases the frequency of particle collisions
(4) increases the activation energy of the reaction   41 _____

42 Given the equation representing a system at equilibrium in a sealed, rigid container:

$$2HI(g) \rightleftharpoons H_2(g) + I_2(g) + energy$$

Increasing the temperature of the system causes the concentration of

(1) HI to increase
(2) H_2 to increase
(3) HI to remain constant
(4) H_2 to remain constant   42 _____

43 Based on Table *I*, which equation represents a reaction with the greatest difference between the potential energy of the products and the potential energy of the reactants?

(1) $4Al(s) + 3O_2(g) \rightarrow 2Al_2O_3(s)$
(2) $2H_2(g) + O_2(g) \rightarrow 2H_2O(\ell)$
(3) $C_3H_8(g) + 5O_2(g) \rightarrow 3CO_2(g) + 4H_2O(\ell)$
(4) $C_6H_{12}O_6(s) + 6O_2(g) \rightarrow 6CO_2(g) + 6H_2O(\ell)$

43 _____

44 Which phase change results in an increase in entropy?

(1) $I_2(g) \rightarrow I_2(s)$ (3) $Br_2(\ell) \rightarrow Br_2(g)$
(2) $CH_4(g) \rightarrow CH_4(\ell)$ (4) $H_2O(\ell) \rightarrow H_2O(s)$

44 _____

45 Given the formula for a compound:

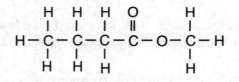

What is the name of this compound?

(1) methyl butanoate   (3) pentanone
(2) methyl butyl ether   (4) pentanoic acid

45 _____

46 Given the equation representing a reaction:

$$2Ca(s) + O_2(g) \rightarrow 2CaO(s)$$

During this reaction, each element changes in
(1) atomic number
(2) oxidation number
(3) number of protons per atom
(4) number of neutrons per atom

46 _____

47 Which equation represents a spontaneous reaction?

(1) $Ca + Ba^{2+} \rightarrow Ca^{2+} + Ba$
(2) $Co + Zn^{2+} \rightarrow Co^{2+} + Zn$
(3) $Fe + Mg^{2+} \rightarrow Fe^{2+} + Mg$
(4) $Mn + Ni^{2+} \rightarrow Mn^{2+} + Ni$

47 _____

48 Which equation represents a neutralization reaction?

(1) $6HClO \rightarrow 4HCl + 2HClO_3$
(2) $CH_4 + 2O_2 \rightarrow CO_2 + 2H_2O$
(3) $Ca(OH)_2 + H_2SO_4 \rightarrow CaSO_4 + 2H_2O$
(4) $Ba(OH)_2 + Cu(NO_3)_2 \rightarrow Ba(NO_3)_2 + Cu(OH)_2$

48 _____

49 Which radioisotope requires long-term storage as the method of disposal, to protect living things from radiation exposure over time?

(1) Pu-239   (3) Fe-53
(2) Fr-220   (4) P-32

49 _____

50 Given the equation representing a reaction:

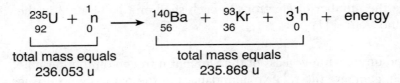

$$^{235}_{92}U + ^{1}_{0}n \longrightarrow ^{140}_{56}Ba + ^{93}_{36}Kr + 3^{1}_{0}n + energy$$

total mass equals
236.053 u

total mass equals
235.868 u

Which statement explains the energy term in this reaction?

(1) Mass is gained due to the conversion of mass to energy.
(2) Mass is gained due to the conversion of energy to mass.
(3) Mass is lost due to the conversion of mass to energy.
(4) Mass is lost due to the conversion of energy to mass.

50 _____

PART B–2

Answer all questions in this part.

Directions (51–65): Record your answers on the answer sheet provided. Some questions may require the use of the *2011 Edition Reference Tables for Physical Setting/Chemistry*.

Base your answers to questions 51 through 53 on the information below and on your knowledge of chemistry.

The only naturally occurring isotopes of nitrogen are N-14 and N-15.

51 State the number of protons in an atom of N-15.   [1]

52 State the number of electrons in each shell of a N-14 atom in the ground state.   [1]

53 Based on the atomic mass of the element nitrogen on the Periodic Table, compare the relative abundances of the naturally occurring isotopes of nitrogen.   [1]

Base your answers to questions 54 through 56 on the information below and on your knowledge of chemistry.

The melting points and boiling points of five substances at standard pressure are listed on the table below.

Melting Points and Boiling Points of Five Substances

Substance	Melting Point (K)	Boiling Point (K)
HCl	159	188
NO	109	121
F_2	53	85
Br_2	266	332
I_2	387	457

54 Identify the substance in this table that is a liquid at STP.   [1]

55 State, in terms of the strength of intermolecular forces, why I_2 has a higher boiling point than F_2.   [1]

56 State what happens to the potential energy of a sample of $NO(\ell)$ at 121 K as it changes to $NO(g)$ at constant temperature and standard pressure.   [1]

Base your answers to questions 57 through 59 on the information below and on your knowledge of chemistry.

A 100.-gram sample of liquid water is heated from 20.0°C to 50.0°C. Enough $KClO_3(s)$ is dissolved in the sample of water at 50.0°C to form a saturated solution.

57 Using the information on Table B, determine the amount of heat absorbed by the water when the water is heated from 20.0°C to 50.0°C.   [1]

58 Based on Table H, determine the vapor pressure of the water sample at its final temperature.   [1]

59 Based on Table G, determine the mass of $KClO_3(s)$ that must dissolve to make a saturated solution in 100. g of H_2O at 50.0°C.   [1]

Base your answers to questions 60 through 62 on the information below and on your knowledge of chemistry.

The diagram and ionic equation below represent an operating voltaic cell.

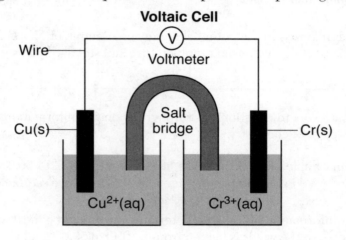

Voltaic Cell

$$3Cu^{2+}(aq) + 2Cr(s) \longrightarrow 3Cu(s) + 2Cr^{3+}(aq)$$

60 Identify the subatomic particles that flow through the wires as the cell operates. [1]

61 State the purpose of the salt bridge in completing the circuit in this cell. [1]

62 Write a balanced equation for the half-reaction that occurs in the copper half-cell when the cell operates. [1]

Base your answers to questions 63 through 65 on the information below and on your knowledge of chemistry.

A NaOH(aq) solution with a pH value of 13 is used to determine the molarity of a HCl(aq) solution. A 10.0-mL sample of the HCl(aq) is exactly neutralized by 16.0 mL of 0.100 M NaOH(aq). During this laboratory activity, appropriate safety equipment was used and safety procedures were followed.

63 Determine the molarity of the HCl(aq) sample, using the titration data. [1]

64 Compare the hydronium ion concentration to the hydroxide ion concentration when the HCl(aq) solution is exactly neutralized by the NaOH(aq) solution. [1]

65 Determine the pH value of a solution that has a H^+(aq) ion concentration 10 times greater than the original NaOH(aq) solution. [1]

PART C

Answer all questions in this part.

Directions (66–85): Record your answers on the answer sheet provided. Some questions may require the use of the *2011 Edition Reference Tables for Physical Setting/Chemistry*.

Base your answers to questions 66 through 68 on the information below and on your knowledge of chemistry.

A hydrate is a compound that has water molecules within its crystal structure. Magnesium sulfate heptahydrate, $MgSO_4 \cdot 7H_2O$, is a hydrated form of magnesium sulfate. The hydrated compound has 7 moles of H_2O for each mole of $MgSO_4$. When 5.06 grams of $MgSO_4 \cdot 7H_2O$ are heated to at least 300.°C in a crucible by using a laboratory burner, the water molecules are released. The sample was heated repeatedly, until the remaining $MgSO_4$ had a constant mass of 2.47 grams. During this laboratory activity, appropriate safety equipment was used and safety procedures were followed.

66 Explain why the sample in the crucible was heated repeatedly until the sample had a constant mass.   [1]

67 Using the lab data, show a numerical setup for calculating the percent composition by mass of water in the hydrated compound.   [1]

68 Determine the gram-formula mass of the magnesium sulfate heptahydrate.   [1]

Base your answers to questions 69 through 71 on the information below and on your knowledge of chemistry.

Solid sodium chloride, also known as table salt, can be obtained by the solar evaporation of seawater and from underground mining. Liquid sodium chloride can be decomposed by electrolysis to produce liquid sodium and chlorine gas, as represented by the equation below.

$$2NaCl(\ell) \rightarrow 2Na(\ell) + Cl_2(g)$$

69 State, in terms of electrons, why the radius of a Na^+ ion in the table salt is smaller than the radius of a Na atom.   [1]

70 Identify the noble gas that has atoms with the same number of electrons as a chloride ion in table salt.   [1]

71 In the space *on your answer sheet*, draw a Lewis electron-dot diagram of a Cl_2 molecule.   [1]

Base your answers to questions 72 through 75 on the information below and on your knowledge of chemistry.

The enclosed cabin of a submarine has a volume of 2.4×10^5 liters, a temperature of 312 K, and a pressure of 116 kPa. As people in the cabin breathe, carbon dioxide gas, $CO_2(g)$, can build up to unsafe levels. Air in the cabin becomes unsafe to breathe when the mass of $CO_2(g)$ in this cabin exceeds 2156 grams.

72 State what happens to the average kinetic energy of the gas molecules if the cabin temperature *decreases*.   [1]

73 Show a numerical setup for calculating the pressure in the submarine cabin if the cabin temperature changes to 293 K.   [1]

74 Determine the number of moles of $CO_2(g)$ in the submarine cabin at which the air becomes unsafe to breathe. The gram-formula mass of CO_2 is 44.0 g/mol.   [1]

75 Convert the original air pressure in the cabin of the submarine to atmospheres.   [1]

Base your answers to questions 76 through 78 on the information below and on your knowledge of chemistry.

Automobile catalytic converters use a platinum catalyst to reduce air pollution by changing emissions such as carbon monoxide, $CO(g)$, into carbon dioxide, $CO_2(g)$. The uncatalyzed reaction is represented by the balanced equation below.

$$2CO(g) + O_2(g) \rightarrow 2CO_2(g) + \text{heat}$$

76 On the labeled axes *on your answer sheet*, draw a potential energy diagram for the reaction represented by this equation.   [1]

77 Compare the activation energy of the catalyzed reaction to the activation energy of the uncatalyzed reaction.   [1]

78 Determine the number of moles of $O_2(g)$ required to completely react with 28 moles of $CO(g)$ during this reaction.   [1]

Base your answers to questions 79 through 81 on the information below and on your knowledge of chemistry.

The solvent 2-chloropropane can be made when chemists react propene with hydrogen chloride, as shown in the equation below.

79 Identify the element in propene that is in all organic compounds.   [1]

80 Explain, in terms of chemical bonds, why the hydrocarbon reactant is classified as unsaturated.   [1]

81 Write the general formula for the homologous series to which propene belongs.   [1]

Base your answers to questions 82 through 85 on the information below and on your knowledge of chemistry.

Radioactive emissions can be detected by a Geiger counter. When radioactive emissions enter the Geiger counter probe, which contains a noble gas such as argon or helium, some of the atoms are ionized. The ionized gas allows for a brief electric current. The current causes the speaker to make a clicking sound. To make sure that the Geiger counter is measuring radiation properly, the device is tested using the radioisotope Cs-137.

To detect gamma radiation, an aluminum shield can be placed over the probe window, to keep alpha and beta radiation from entering the probe. A diagram that represents the Geiger counter is shown below.

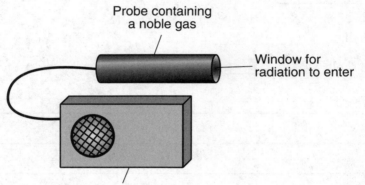

Probe containing
a noble gas

Window for
radiation to enter

Amplifier and speaker

82 Compare the first ionization energy of argon to the first ionization energy of helium. [1]

83 State evidence from the passage that gamma radiation has greater penetrating power than alpha or beta radiation. [1]

84 Determine the time required for a sample of cesium-137 to decay until only $\frac{1}{8}$ of the original sample remains unchanged. [1]

85 Complete the nuclear equation *on your answer sheet* for the decay of Cs-137 by writing a notation for the missing product. [1]

Answer Sheet
August 2019
Chemistry—Physical Setting

PART B–2

51 _____

52 First shell: _____

Second shell: _____

53 _____

_____

_____

54 _____

55 _____

_____

_____

56 _____

_____

_____

57 _____ J

58 _____ kPa

59 _____ g

60 _____

61 _____

_____

_____

62 _____

63 _____ M

64 _____

_____

_____

65 _____

PART C

66 _____

_____

_____

67

68 _____ g/mol

69 _____

_____

_____

70 _____

71

72 _____

_____

_____

73 _____

74 _____ mol

75 _____ atm

76

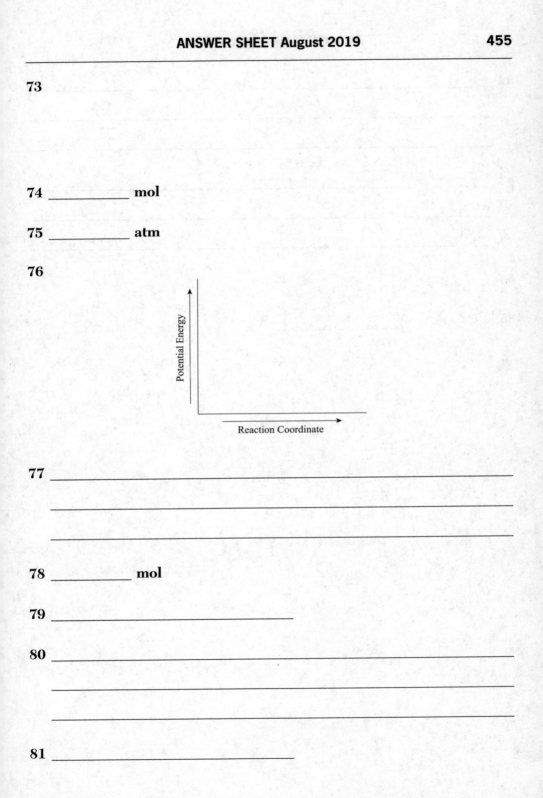

Potential Energy

Reaction Coordinate

77 _____

_____

_____

78 _____ mol

79 _____

80 _____

_____

_____

81 _____

82 _____

_____

_____

83 _____

_____

_____

84 _____ y

85 $^{137}_{55}\text{Cs} \rightarrow \,^{0}_{-1}\text{e} +$ _____

Answers
August 2019
Chemistry—Physical Setting

Answer Key

PART A

1. 1	7. 1	13. 4	19. 1	25. 1
2. 1	8. 4	14. 1	20. 3	26. 4
3. 4	9. 3	15. 1	21. 3	27. 4
4. 2	10. 3	16. 2	22. 2	28. 3
5. 4	11. 2	17. 1	23. 3	29. 4
6. 2	12. 4	18. 4	24. 3	30. 3

PART B–1

31. 2	35. 4	39. 3	43. 1	47. 4
32. 2	36. 1	40. 2	44. 3	48. 3
33. 3	37. 4	41. 3	45. 1	49. 1
34. 1	38. 3	42. 1	46. 2	50. 3

PART B–2 and **PART C**. *See* **Answers Explained**.

Answers Explained

PART A

1. **1** The idea that atoms were the smallest particles of matter began with the ancient Greeks in the 5th century B.C.E. Credit is given to Democritus and Leucippus. These philosophers imagined atoms to be tiny, hard, indivisible spheres of matter.

WRONG CHOICES EXPLAINED:
(2) Ernest Rutherford developed the nuclear model of the atom in 1911 as a result of his famous gold foil experiment.

(3) J. J. Thompson discovered the presence of negatively charged electrons in atoms in 1897 as a result of his famous cathode ray tube experiment.

(4) Wave-particle duality is a cornerstone premise in the quantum model of the atom. Louis de Broglie, a French physicist, first proposed this concept in 1923.

2. **1** All atoms are electrically neutral. The number of negatively charged electrons must equal the number of positively charged protons.

3. **4** Protons and neutrons each have a mass of 1 amu (atomic mass unit). Reference Table O contains information about nuclear particles. Check to verify that protons and neutrons have the same mass (the upper number on each symbol).

4. **2** This is a definition question. Know this definition! The modern quantum model of the atom places electrons in orbitals of varying energies around the nucleus.

5. **4** Electron shells closer to the nucleus are lower in energy than electron shells farther from the nucleus. This is because negatively charged electrons are attracted to the positively charged nucleus.

6. **2** Electronegativity is a measure of an atom's attraction for electrons and specifically for a bonded pair of electrons. Electronegativity increases across any period of the Periodic Table of the Elements because of the increasing nuclear charge of each successive element. Use Reference Table S for any questions involving periodic trends! Using the electronegativity data for elements in Period 2, you can see that the values increase as elements are considered from left to right across the period.

WRONG CHOICES EXPLAINED:

(1) Look at the atomic radius data shown on Reference Table S. The atomic radius decreases as elements in Period 2 are considered from lithium to fluorine. As nuclear charge (the number of protons in the nucleus) increases, the atomic radius decreases.

(3) Look at the electron configurations on the Periodic Table of the Elements. The electrons in each element in Period 2 occupy the first and second electron shell.

(4) There are 2 electrons in the first electron shell of every atom in the ground state except for hydrogen, which has only 1 electron.

7. **1** Six elements are classified as metalloids, and these must be memorized! Metalloids straddle the dividing line between metals and nonmetals. They exhibit properties of both metals and nonmetals. The six metalloid elements are boron (B), silicon (Si), arsenic (As), tellurium (Te), antimony (Sb), and germanium (Ge).

WRONG CHOICES EXPLAINED:

(2) Potassium (K) is an alkali metal (Group 1).

(3) Sulfur (S) is a nonmetal.

(4) Xenon (Xe) is a noble gas.

8. **4** Strontium and barium are both in Group 2 on the Periodic Table of the Elements, and they both have 2 valence electrons. Valence electrons are involved in forming chemical bonds between atoms and are therefore largely responsible for an atom's chemical properties.

9. **3** The chemical formula for a compound, for example H_2O, describes the exact proportion of elements in the compound. In H_2O, there are 2 hydrogen atoms for every oxygen atom.

10. **3** Chemical formulas provide information about the types and proportion of elements in a compound. The molecular formula provides the number of atoms of each element present in 1 molecule of a specific compound. The structural formula shows the arrangement of atoms in the compound. Consider the example of butane shown below:

Molecular Formula	Structural Formula

$$C_4H_{10}$$

$$\begin{array}{ccccc} H & H & H & H \\ | & | & | & | \\ H-C-C-C-C-H \\ | & | & | & | \\ H & H & H & H \end{array}$$

11. **2** Elements classified as metals are held together by metallic bonds. All metals are found on the left of the staircase dividing line on the Periodic Table of the Elements. Cesium (Cs) is an alkali metal in Group 1.

WRONG CHOICES EXPLAINED:
(1) Bromine (Br) is a nonmetal in Group 17.
(3) Krypton (Kr) is a noble gas in Group 18.
(4) Sulfur (S) is a nonmetal in Group 16.

12. **4** Draw a Lewis structure to show how many covalent bonds are shared between nitrogen atoms in a nitrogen diatomic molecule. The Lewis structure for nitrogen, N_2, is shown below:

$$:N{\equiv}N:$$

Since the nitrogen atoms form 3 covalent bonds and since 2 electrons are shared in each covalent bond, there are 6 electrons shared between nitrogen atoms.

13. **4** The reaction shows two independent hydrogen atoms reacting to form a stable diatomic hydrogen molecule. The formation of chemical bonds releases energy since atoms are more stable once they are bonded together.

14. **1** To answer this question correctly, you need to recall two things: (1) electronegativity is defined as an atom's attraction for electrons in a covalent bond, and (2) values for electronegativity can be found on Reference Table *S*. Looking at Reference Table *S*, we find that chlorine's electronegativity is greater than that of carbon, phosphorus, or sulfur.

15. **1** Chemical properties involve the chemical reactivity of a substance, and this depends on the bonding and chemical composition within the substance. Physical changes to a substance, like having more or less mass, do not affect chemical properties.

WRONG CHOICES EXPLAINED:
(2) The 100.-gram sample of $H_2O(\ell)$ has a larger volume than a 50.-gram sample of $H_2O(\ell)$.
(3) Standard temperature is 0°C, and both samples are at STP.
(4) Both samples are made of H_2O. So, both samples have the same chemical formula and also the same empirical formula.

16. **2** Sand does not dissolve in water. When the two are mixed together, the sand will settle to the bottom since it is more dense than water. Since a sand/water mixture has a layer of solid sand and a layer of liquid water, it is a heterogeneous mixture in two different phases.

17. **1** Distillation is the laboratory process used to separate liquid mixtures based on boiling point. As the liquid mixture begins to boil, the component of the mixture with the lower boiling point will boil away first. By passing the vapors through a condenser, the components of the mixture can be collected separately.

18. **4** One of the principles of the kinetic molecular theory of gases is that since gas particles are so small, the volume occupied by the particles themselves is considered to be negligible compared with the volume of their container.

WRONG CHOICES EXPLAINED:
    (1) The kinetic molecular theory states that gas particles move in random, constant, straight-line motion.
    (2) The kinetic molecular theory states that gas particles undergo perfectly elastic collisions in which the total energy of the system remains the same before and after collisions.
    (3) The kinetic molecular theory states that gas particles have no attractions or repulsions for each other.

19. **1** The gas phase is far less dense than any of the other phases. Because gas particles are extremely tiny and because the distances between gas particles are very large relative to their size, we cannot see gases. We can see liquids, solids, and solutions because their particles are much closer together.

20. **3** At equilibrium, the rate of the forward reaction must equal the rate of the reverse reaction. Since both products and reactants are forming at equal rates, the concentration of each remains constant. A common misconception is that equilibrium involves equal amounts of reactants and products, but this is not the case! Equilibrium is about equal rates, not equal amounts.

21. **3** Systems naturally tend to undergo changes that produce lower total energy and greater disorder or randomness (entropy). For example, when leaves fall from a tree, they fall to the ground (a position of lower potential energy) and scatter in many directions (greater randomness).

22. **2** Alkenes and alkynes are two homologous series of hydrocarbons (see Reference Table Q). All hydrocarbons contain just two elements: carbon and hydrogen.

23. **3** Reference Table R lists the general formulas of common organic functional groups. Only amides contain both nitrogen and oxygen.

24. **3** Electrons are tiny charged particles of matter. The moles of electrons lost by $Mg(s)$ must be equal to the number of moles of electrons gained by the $O_2(g)$. This conservation of charge is present during all chemical reactions.

25. **1** The anode is defined as the electrode where oxidation takes place, and the cathode is defined as the electrode where reduction takes place. The mnemonic device "An Ox, Red Cat" can help you remember these definitions!

26. **4** K_2SO_4 and $MgSO_4$ are both electrolytes, meaning that they dissociate completely into freely moving ions in aqueous solution. Freely moving dissociated ions allow charge to flow through the solution. However, K_2SO_4 will dissociate into *more* ions when it dissociates, and the more ions in solution, the greater the solution's capacity to conduct electricity.

WRONG CHOICES EXPLAINED:
(1), (2) These compounds are molecular compounds. They do not dissociate into ions in aqueous solution, and therefore they are nonelectrolytes.
(3) $MgSO_4$ dissociates into fewer ions than does K_2SO_4 when placed into an aqueous solution. Therefore, $MgSO_4$ is less effective as an electrolyte than K_2SO_4.

27. **4** Reference Table M lists the colors of common acid-base indicators at different pH levels. Bromcresol green is yellow at all pH values less than 3.8.

WRONG CHOICES EXPLAINED:
(1) Bromcresol green is blue at pH values greater than 5.4.
(2) Bromcresol green is green at pH values between 3.8 and 5.4.
(3) Bromcresol green is never red.

28. **3** $H_3O^+(aq)$ is the hydronium ion. Hydronium ions form when hydrogen ions (H^+) are released into water. The H^+ protonates the water molecules, resulting in hydronium ions.

29. **4** Both natural and artificial transmutation reactions are nuclear reactions that convert atoms of one element into atoms of another element. Natural transmutation, also known as radioactive decay, involves the spontaneous disintegration of an unstable nucleus. Artificial transmutation reactions take place when a high-speed particle collides with a target nucleus.

30. **3** Nuclear fusion happens when two lighter weight nuclei merge to form heavier nuclei. An equation representing a nuclear fusion reaction is shown below:

$$\,_1^2\mathrm{H} + \,_1^3\mathrm{H} \rightarrow \,_2^4\mathrm{He} + \,_0^1\mathrm{n}$$

WRONG CHOICES EXPLAINED:

(1), (2) Both alpha decay and beta decay are types of natural transmutation reactions.

(4) Fission reactions occur when a high-speed neutron collides with a U-235 nucleus, forcing the nucleus to split into two or more smaller nuclei. The formation of neutrons enables a chain reaction to occur.

(NOTE: You need to be able to recognize the four types of nuclear reactions. These include natural transmutation, artificial transmutation, fission, and fusion. Memorize these reactions!)

PART B–1

31. **2** All elements produce unique bright-line spectra when excited electrons in these atoms fall back to the ground state. The bright-line spectrum of a mixture of two or more elements contains all of the wavelength bands of each separate element superimposed together on the same spectrum. Element X can quickly be eliminated since the two wavelength bands at ~740 nm are not evident in the mixture. Likewise, element D can be eliminated because the two wavelength bands at ~590 nm are not present in the mixture. Careful inspection shows that all wavelength bands from elements A and Z are found in the mixture.

32. **2** The Periodic Table of the Elements in the Chemistry Reference Tables shows the ground state configuration of sulfur as 2–8–6. A sulfur atom would be in an excited state if its configuration were 2–7–7 because this configuration shows that a second shell electron has moved to the third shell.

WRONG CHOICES EXPLAINED:

(1) This is the ground state electron configuration for a sulfur atom. The question asks for an *excited* state configuration.

(3) 2–8–7 is the ground state configuration for an atom of chlorine.

(4) Sulfur atoms have 16 electrons. The electron configuration 2–7–8 has a total of 17 electrons and is an excited state electron configuration for a chlorine atom.

33. **3** Atoms with the same number of protons and different numbers of neutrons are isotopes of one another. Cl-35 and Cl-37 are both atoms of chlorine. So, they both have an atomic number of 17, and therefore each has 17 protons. The mass numbers differ because Cl-37 has 2 more neutrons than does Cl-35.

WRONG CHOICES EXPLAINED:

(1) Hydrogen (H) and helium (He) are different elements and have different numbers of protons.

(2) S-32 is identical to S-32. These atoms have the same number of protons and the same number of neutrons.

(4) Gallium (Ga) and germanium (Ge) are different elements and have different numbers of protons.

34. **1** NH_4SCN is an ionic compound formed from two polyatomic ions listed on Reference Table *E*. NH_4^+ is the ammonium ion, and SCN^- is the thiocyanate ion.

35. **4** Conservation of atoms means that the number of atoms of each type of element must be the same before and after the reaction. In answer choice (4), there are 4 iron (Fe) atoms and 6 oxygen (O) atoms on each side of the equation.

36. **1** Covalent bonds form between nonmetal atoms, such as between hydrogen (H) and oxygen (O). When metal atoms bond with nonmetal atoms, ionic bonds form.

37. **4** Oxygen is a diatomic gas at STP.

WRONG CHOICES EXPLAINED:

(1) This model shows a monatomic gas, like one of the noble gases.

(2) This model shows the regular arrangement and tight packing of a solid.

(3) This model does not show the large spaces between particles that exist in samples of gas.

38. **3** Real gases behave more like ideal gases when temperature is high and pressure is low. These conditions minimize any possible attractions between gas particles, allowing the gas to adhere more closely to the principles of the kinetic molecular theory of gases. A good way to remember this is to think of summertime, when the temperature is high and the pressure is low!

39. **3** The molarity formula is found on Reference Table T:

$$\text{molarity} = \frac{\text{moles of solute}}{\text{liter of solution}}$$

Plug the given information into this equation to solve for the molarity:

$$\text{molarity} = \frac{0.50 \text{ mol}}{2.0 \text{ L}} = 0.25 \text{ M}$$

40. **2** Equal volumes of gas at the same conditions of temperature and pressure contain the same number of particles. This relationship is known as Avogadro's law. For example, the molar volume of a gas at STP equals 22.4 L/mol no matter what type of gas you have. Since gas samples 1 and 3 have equal volumes and are at the same temperature and pressure, they contain the same number of particles.

41. **3** Reactions occur only when reactant particles collide. The powdered zinc exposes more zinc atoms to collisions with HCl molecules.

42. **1** Le Châtelier's principle states that chemical systems at equilibrium that experience a stress will respond to restore equilibrium. If temperature is increased in this equilibrium system, the reaction will respond by favoring the reverse reaction to use up the extra energy and bring the temperature back down again, restoring equilibrium. This "shift to the left" causes more HI(g) to be produced.

43. **1** The heat of reaction, ΔH, is defined as the difference between the potential energy of the products in a chemical reaction and the potential energy of the reactants. Know this definition! Reference Table I lists the heats of reaction for various chemical reactions. Use this table to find the largest value of the heat of reaction for the equations shown. Since you are only looking for the difference between the potential energy of the products and the potential energy of the reactants, the sign on the ΔH does not matter. The difference is greatest for $4\text{Al}(s) + 3\text{O}_2(g) \rightarrow 2\text{Al}_2\text{O}_3(s)$, with a ΔH of -3351 kJ.

44. **3** Entropy is defined as the degree of chaos, randomness, or disorder within a system or a substance. Increasing entropy is one important factor that leads to spontaneous chemical or physical change. Phase changes that lead to a less orderly arrangement of atoms increase the entropy of a sample of matter. Particles in a gas are far less ordered than particles in a liquid.

WRONG CHOICES EXPLAINED:
(1) A gas to solid change results in a much more ordered arrangement of atoms. This is the opposite of entropy!
(2) A gas to liquid change increases the order (and decreases the entropy) in a sample of matter. Liquid particles have far greater attractions to one another than do gas particles, and their motion is less random than the motion of gas particles.
(4) A liquid to solid change results in a much more ordered arrangement of atoms. This is the opposite of entropy!

45. **1** To name the compound in this problem (and shown below), classify the compound according to which organic functional group it contains. Organic functional groups are listed in Reference Table *R*:

This compound contains the ester functional group, and the name will end with the suffix *–oate*. There is only one answer choice with the correct suffix: choice (1).

46. **2** This synthesis does not affect the nuclei of the atoms involved. However, before the reaction, Ca(s) has an oxidation number of 0. After the reaction, the Ca^{2+} ion has an oxidation number of +2. Oxygen also changes in oxidation number, from 0 before the reaction to −2 after the reaction.

47. **4** Reference Table *J* organizes metals from the most active metals at the top to the least active metals at the bottom. Metal atoms oxidize when they react, and they can give electrons to the ions of less reactive metals. For a spontaneous reaction, a pure metal must be paired with the ion of a metal that is lower on Reference Table *J*.

WRONG CHOICES EXPLAINED:

(1), (2), (3) The pure metals in these reactions are located below the ions with which they are paired. This means that the pure metals are less reactive than the ions, and consequently, the pure metals will not be able to oxidize (no reaction will happen).

48. **3** The reaction between $Ca(OH)_2$ and H_2SO_4 is a neutralization reaction. Neutralization reactions are really double replacement reactions between acids and bases. The word equation for neutralization reactions between Arrhenius acids and bases is the following:

$$ACID + BASE \rightarrow WATER + SALT$$

$Ca(OH)_2$ is a base listed on Reference Table L, and H_2SO_4 is an acid listed on Reference Table K.

WRONG CHOICES EXPLAINED:

(1) This reaction is an example of a decomposition reaction.

(2) This reaction is an example of a combustion reaction.

(4) This reaction is an example of a double replacement reaction between two ionic compounds.

49. **1** Reference Table N shows that Pu-239 has a half-life of 2.410×10^4 years. With a half-life this long, this radioisotope would continue to emit radiation for thousands of years. The other radioisotopes have half-lives on the order of seconds, minutes, or days and would not remain radioactive for very long.

50. **3** Nuclear fission is depicted in the equation for this question. Nuclear fission produces energy by converting a small amount of mass into energy. The amount of energy produced can be calculated by Einstein's famous equation $E = mc^2$, in which m is the mass lost in the fission reaction and c is the speed of light.

PART B–2

[All questions in Part B–2 are worth 1 point.]

51. All nitrogen atoms have **7** protons. Nitrogen's atomic number is 7, which equals the number of protons in the nucleus of every nitrogen atom.

52. The Periodic Table of the Elements shows that the electron configuration for the element nitrogen is 2–5. This means that **2 electrons are in the first electron shell and 5 electrons are in the second electron shell**.

53. The atomic mass of an element is the weighted average of the masses of the naturally occurring isotopes of that element. The atomic mass of nitrogen as shown on the Periodic Table of the Elements is 14.0067 amu. Since this weighted average is much closer to 14 than to 15, there must be a **far greater amount of N-14 than N-15**.

54. **Bromine, Br_2**, is a liquid because standard temperature, 273 K, lies between bromine's melting point and boiling point.

55. **Iodine, I_2, has stronger intermolecular forces than fluorine, F_2.** Stronger intermolecular forces between molecules in a liquid sample cause higher boiling points. More energy is required to allow molecules to overcome the forces of attraction between them and separate completely to form a gas.

56. **Potential energy increases** as NO changes phase from a liquid to a gas at 121 K. Potential energy is stored energy. Particles gain potential energy as they move farther apart to form a gas.

57. Heat formulas are found on Reference Table T. Since the temperature of the water increases from 20.0°C to 50.0°C, use the following formula:

$$q = mC\Delta T$$

In the formula above, q is the amount of heat in joules (J) lost or gained, m is the mass of the water in grams, and ΔT is the change in temperature in Kelvin. To determine the change in temperature, remember that 1 degree Celsius equals 1 Kelvin and calculate the following:

$$\Delta T = T_{final} - T_{initial}$$
$$= 50.0°C - 20.0°C$$
$$= 30.0°C$$
$$= 30.0 \text{ K}$$

Using the value for the specific heat capacity shown on Reference Table B and plugging in the given values leads to the following expression:

$$q = (100. \text{ g})(4.18 \text{ J/g·K})(30.0 \text{ K}) \cong \textbf{12,500 J}$$

(NOTE: All values from **12,500 J to 13,000 J** were accepted as correct answers to this problem.)

58. The vapor pressure of water at 50.0°C is **12 kPa**. Read this value directly from Reference Table *H*.

(NOTE: All values from **11 kPa to 13 kPa** were accepted as correct answers to this problem.)

59. **22 g** of $KClO_3$ must be added to 100. g of water at 50.0°C to make a saturated solution. Read this value directly from Reference Table *G*.

(NOTE: All values from **20 g to 23 g** were accepted as correct answers to this problem.)

60. **Electrons** are lost as oxidation occurs at the anode. These electrons travel through the wire to the cathode, where they are gained during the process of reduction.

61. **The salt bridge allows ions to flow into each compartment** in the voltaic cell **to maintain electrical neutrality** in the system. A buildup of positive or negative charges in either compartment would cause electrons to stop flowing.

62. Copper is below chromium on Reference Table *J*, so copper ions reduce in the presence of chromium. Reduction is defined as the "gain of electrons." Copper ions gain 2 electrons to form neutral copper atoms, Cu^0, as shown in the following reduction half-cell equation:

$$Cu^{2+}(aq) + 2e^- \rightarrow Cu^0(s)$$

(NOTE: Including state symbols is not required to earn credit.)

63. Titration is the lab technique used to determine the concentration of an acid or base by neutralizing it with an acid or base of known concentration. The titration formula is found on Reference Table *T*:

$$M_A V_A = M_B V_B$$

In this formula, M_A is the molarity of H^+, V_A is the volume of acid, M_B is the molarity of OH^-, and V_B is the volume of base.

Substitute the values given in the problem and rearrange to solve:

$$M_{(A)}(10.0 \text{ mL}) = (0.100 \text{ M})(16.0 \text{ mL})$$

$$M_A = \frac{(0.100 \text{ M})(16.0 \text{ mL})}{10.0 \text{ mL}}$$

$$= \mathbf{0.160 \text{ M}}$$

64. At the point of neutralization, **the concentration of hydronium ions is exactly equal to the concentration of hydroxide ions**. The net ionic equation for neutralization is $H^+(aq) + OH^-(aq) \rightarrow H_2O(\ell)$. Since H^+ and OH^- are present in a 1:1 molar relationship in this equation, these ions must be present in equal amounts for neutralization to occur. Remember that $H^+(aq)$ is equivalent to a hydronium ion $H_3O^+(aq)$.

65. pH measures the concentration of hydrogen ion, H^+, in aqueous solution. pH is a logarithmic function, meaning that it is based on the exponent of the base 10 molarity of the hydrogen ion content. Each pH value is 10× different in hydrogen ion molarity than the pH immediately higher and lower than itself. Hydrogen ions make a solution acidic, and solutions that are more acidic have lower pH values. The problem states that the original NaOH solution has a pH of 13. The pH of a solution that has 10 times *more* $H^+(aq)$ ion is one number *lower* than the original pH—in other words, **12**.

PART C

[All questions in Part C are worth 1 point.]

66. The reading passage for this question states that when the magnesium sulfate heptahydrate $(MgSO_4 \cdot 7H_2O)$ is heated to a high temperature in a crucible, water molecules are released from the compound. **The sample was heated repeatedly to ensure that all of the water in the sample was removed.**

67. The formula for percent composition is found on Reference Table *T*. In this problem, you are specifically asked to solve for the percent composition by mass of water in the hydrate:

$$\% \text{ composition by mass of water} = \frac{\text{mass of water}}{\text{total mass of hydrate}} \times 100$$

The mass of water can be found by subtracting the dry solid mass left in the crucible after heating from the initial mass of the hydrate that was placed into the crucible. The difference between these masses is the mass of the water that was driven off during the heating process:

$$5.06 \text{ g hydrate} - 2.47 \text{ g anhydrous residue} = 2.59 \text{ g water}$$

Plugging this value into the equation above:

$$\% \text{ composition by mass of water} = \frac{2.59 \text{ g water}}{5.06 \text{ g hydrate}} \times 100$$

(NOTE: This question asks ONLY for the numerical setup for calculating the mass percent of water in the hydrate. The numerical setup is all that you should record on your answer sheet. If you would like to check your work by solving for mass percent of water, do it on scrap paper.)

68. Calculate the gram-formula mass, also known as the molar mass, by adding the atomic masses of each element in the compound. The molar mass of each element is found on the Periodic Table of the Elements:

Calculating Gram-Formula Mass of $MgSO_4 \cdot 7H_2O$			
Element	**Atomic Mass**	**Number of Atoms**	**Total Mass of Element**
Mg	24.3 g/mol	1	24.3 g/mol
S	32.1 g/mol	1	32.1 g/mol
O	16.0 g/mol	4	64.0 g/mol
H	1.0 g/mol	14	14.0 g/mol
O	16.0 g/mol	7	112.0 g/mol
		Gram-Formula Mass	**246.4 g/mol**

(NOTE: Acceptable values range from **245.989 g/mol to 247 g/mol**.)

69. The electron configuration of the element sodium is 2–8–1. When the Na^+ ion forms, the outmost electron is lost, so the sodium ion has an electron configuration of 2–8. The radius of the Na^+ ion is smaller than the radius of a Na atom

because **the sodium atom has an electron in the third electron shell whereas the sodium ion only has electrons in the first and second electron shells**.

70. The chloride ion, Cl^-, forms when an atom of chlorine gains 1 electron, giving the ion a total of 18 electrons (which is 1 more than an atom of chlorine). Atoms of the noble gas **argon** also contain 18 electrons.

71. Lewis electron-dot diagrams show the distribution of valence electrons within a molecule. Lewis electron-dot diagrams help to illustrate the number and types of bonds in a compound. Each Cl atom has 7 valence electrons. By sharing one pair of electrons in a single covalent bond, each chlorine atom achieves a stable octet of valence electrons:

$$:\ddot{C}l - \ddot{C}l:$$

72. Temperature is directly related to the kinetic energy of particles in a sample of matter. Kinetic energy is the energy of particle motion. Putting these two ideas together leads to the understanding that particles in a sample of matter move faster at higher temperatures and slower at lower temperatures. If the cabin temperature decreases, **the average kinetic energy of the gas molecules would also decrease**.

73. Use the combined gas law equation on Reference Table T to solve this problem:

$$\frac{P_1 V_1}{T_1} = \frac{P_2 V_2}{T_2}$$

Since the volume of the submarine cabin does not change, $V_1 = V_2$, and we can simplify the formula to the following:

$$\frac{P_1}{T_1} = \frac{P_2}{T_2}$$

Plug the given information into the equation:

$$\frac{116 \text{ kPa}}{312 \text{ K}} = \frac{P_2}{293 \text{ K}}$$

(NOTE: This question asks ONLY for the numerical setup for calculating the pressure in the submarine cabin. The numerical setup is all that you should record on your answer sheet. If you would like to check your work by solving for the pressure, do it on scrap paper.)

74. The introductory paragraph to this question states, "Air in the cabin becomes unsafe to breathe when the mass of $CO_2(g)$ in this cabin exceeds 2156 grams." To convert this mass into moles of $CO_2(g)$, use the mole calculation formula found on Reference Table T:

$$\text{number of moles} = \frac{\text{given mass}}{\text{gram-formula mass}}$$

Plugging in the numbers:

$$\text{number of moles} = \frac{2156 \text{ g } CO_2}{44.0 \text{ g/mol } CO_2} = \textbf{49.0 mol } CO_2$$

(NOTE: All values from **49 mol to 50 mol** of CO_2 were accepted as correct answers to this problem.)

75. The relationship between kPa and atm is given on Reference Table A: 101.3 kPa = 1 atm. The conversion from the given pressure in kPa to pressure in atm can be accomplished using either dimensional analysis or ratios.

Using dimensional analysis:

$$116 \text{ } \cancel{\text{kPa}} \times \frac{1 \text{ atm}}{101.3 \text{ } \cancel{\text{kPa}}} = 1.15 \text{ atm}$$

Using ratios:

$$\frac{1 \text{ atm}}{101.3 \text{ kPa}} = \frac{x \text{ atm}}{116 \text{ kPa}}$$

$$x = \textbf{1.15 atm}$$

(NOTE: All values from **1.14 atm to 1.16 atm** were accepted as correct answers to this problem.)

76. The equation given for the conversion of CO(g) to CO$_2$(g) shows the heat term on the product side of the reaction. Since heat is produced, this is an exothermic reaction. Some of the potential energy of the reactant is lost during this reaction as a lower energy product is produced, as shown in the potential energy diagram below:

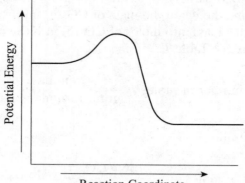

77. Adding a catalyst to a chemical reaction lowers the activation energy (E_a) for the reaction, making it possible for more collisions to have sufficient energy for particles to react. **The catalyzed reaction has a lower activation energy than the uncatalyzed reaction.**

78. All coefficients in chemical reactions represent particles or moles. Mole ratios can be used to find the number of moles of any substance in a chemical equation if the moles of another substance are known. Solve this stoichiometry problem using either dimensional analysis or ratios.

Using dimensional analysis:

$$\frac{28 \text{ mol CO(g)}}{1} \times \frac{1 \text{ mol O}_2(g)}{2 \text{ mol CO(g)}} = 14 \text{ mol O}_2(g)$$

Using mole ratios:

$$\frac{28 \text{ mol CO(g)}}{2} = \frac{x \text{ mol O}_2(g)}{1}$$

$$x = \textbf{14 mol O}_2\textbf{(g)}$$

79. Organic chemistry, by definition, is the chemistry of **carbon**-based compounds. Carbon atoms form 4 covalent bonds and can easily bond with other carbon atoms to form chains, rings, and networks.

80. Unsaturated hydrocarbons contain at least one double or triple bond between carbon atoms. The term "unsaturated" implies that other atoms could be added to the compound by breaking the double or triple bond between carbon atoms. **The hydrocarbon reactant is classified as an unsaturated hydrocarbon because of the double bond between the carbon atoms.**

81. Propene belongs to the alkene family of hydrocarbons. The general formula for alkenes, shown on Reference Table Q, is $\mathbf{C_n H_{2n}}$.

82. First ionization energy data is listed on Reference Table S. **Argon has a first ionization energy of 1521 kJ/mol, while helium has a first ionization energy of 2372 kJ/mol.**

83. The passage states that an aluminum shield can be placed over the probe window, preventing alpha and beta radiation from entering the probe. The passage says that gamma radiation can still be measured when the aluminum shield is over the probe window. Therefore, **gamma radiation has greater penetrating power than either alpha or beta radiation because gamma radiation can pass through aluminum, but alpha and beta radiation cannot pass through aluminum**.

84. From Reference Table N, the half-life of Cs-137 is 30.2 years. One half-life is the time required for a radioisotope to decay to half of its original mass. After one half-life, only half of the original Cs-137 remains. After another half-life passes, only half of this half remains. This continues for each half-life:

$$\frac{1}{2} \times \frac{1}{2} \times \frac{1}{2} = \frac{1}{8}$$

The process of radioactive decay is represented below:

| Initial Amount 100% | After one half-life, $\frac{1}{2}$ remains | After 2nd half-life, $\frac{1}{4}$ remains | After 3rd half-life, $\frac{1}{8}$ remains |

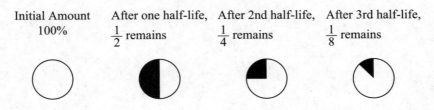

Three half-lives must pass for Cs-137 to decay until only $\frac{1}{8}$ of the original sample remains unchanged. Since each half-life is 30.2 years long, calculate the total time elapsed:

$$3 \times 30.2 \text{ years} = \textbf{90.6 years}$$

85. To balance a nuclear equation, the sum of the atomic numbers (bottom) and the mass numbers (top) must be equal on both sides of the arrow. The missing product must have a mass number of 137 since the mass of a beta particle is 0. The atomic number of the missing product must be 56 since $56 - 1 = 55$. The missing product is an atom of the element barium, Ba, since barium has an atomic number of 56. The missing product is therefore $^{137}_{56}\textbf{Ba}$.

(NOTE: **Ba-137**, **barium-137**, and ^{137}**Ba** are all acceptable responses.)

Mark (✓) the questions you answered correctly. Count the number of checks and follow the formulas given to determine your score on each topic.

Core Area	☐ Questions Answered Correctly

75: (1)

Section M—Math Skills
☐ Number of checks ÷ 1 × 100 = ___ %

66, 82: (2)

Section R—Reading Skills
☐ Number of checks ÷ 2 × 100 = ___ %

1–5, 31–33, 52, 53: (10)

Section I—Atomic Concepts
☐ Number of checks ÷ 10 × 100 = ___ %

6–8, 51: (4)

Section II—Periodic Table
☐ Number of checks ÷ 4 × 100 = ___ %

9, 10, 34, 35, 67, 68, 74, 78: (8)

Section III—Moles/Stoichiometry
☐ Number of checks ÷ 8 × 100 = ___ %

11–14, 36, 69–71: (8)

Section IV—Chemical Bonding
☐ Number of checks ÷ 8 × 100 = ___ %

15–19, 37–40, 54–59, 72, 73: (17)

Section V—Physical Behavior of Matter
☐ Number of checks ÷ 17 × 100 = ___ %

20, 21, 41–44, 76, 77: (8)

Section VI—Kinetics and Equilibrium
☐ Number of checks ÷ 8 × 100 = ___ %

Core Area □ Questions Answered Correctly

22, 23, 45, 79–81: (6)

Section VII—Organic Chemistry
□ Number of checks ÷ 6 × 100 = ___ %

24, 25, 46, 47, 60–62: (7)

Section VIII—Oxidation–Reduction
□ Number of checks ÷ 7 × 100 = ___ %

26–28, 48, 63–65: (7)

Section IX—Acids, Bases, and Salts
□ Number of checks ÷ 7 × 100 = ___ %

29, 30, 49, 50, 83–85: (7)

Section X—Nuclear Chemistry
□ Number of checks ÷ 7 × 100 = ___ %

NOTES

NOTES

NOTES

NOTES

NOTES

NOTES